SpringerBriefs in Molecular Science

For further volumes:
http://www.springer.com/series/8898

Silvia A. Brandán

A Structural and Vibrational Study of the Chromyl Chlorosulfate, Fluorosulfate, and Nitrate Compounds

 Springer

Silvia A. Brandán
Facultad de Bioquímica
 Instituto de Química Inorgánica
Universidad Nacional de Tucumán
San Miguel de Tucumán, Tucumán
Argentina

ISSN 2191-5407   ISSN 2191-5415 (electronic)
ISBN 978-94-007-5762-2   ISBN 978-94-007-5763-9   (eBook)
DOI 10.1007/978-94-007-5763-9
Springer Dordrecht Heidelberg New York London

Library of Congress Control Number: 2012951395

© The Author(s) 2013
This work is subject to copyright. All rights are reserved by the Publisher, whether the whole or part of the material is concerned, specifically the rights of translation, reprinting, reuse of illustrations, recitation, broadcasting, reproduction on microfilms or in any other physical way, and transmission or information storage and retrieval, electronic adaptation, computer software, or by similar or dissimilar methodology now known or hereafter developed. Exempted from this legal reservation are brief excerpts in connection with reviews or scholarly analysis or material supplied specifically for the purpose of being entered and executed on a computer system, for exclusive use by the purchaser of the work. Duplication of this publication or parts thereof is permitted only under the provisions of the Copyright Law of the Publisher's location, in its current version, and permission for use must always be obtained from Springer. Permissions for use may be obtained through RightsLink at the Copyright Clearance Center. Violations are liable to prosecution under the respective Copyright Law.
The use of general descriptive names, registered names, trademarks, service marks, etc. in this publication does not imply, even in the absence of a specific statement, that such names are exempt from the relevant protective laws and regulations and therefore free for general use.
While the advice and information in this book are believed to be true and accurate at the date of publication, neither the authors nor the editors nor the publisher can accept any legal responsibility for any errors or omissions that may be made. The publisher makes no warranty, express or implied, with respect to the material contained herein.

Printed on acid-free paper

Springer is part of Springer Science+Business Media (www.springer.com)

Contents

Contributors

Silvia A. Brandán Facultad de Bioquímica, Química y Farmacia, Instituto de Química Inorgánica, Universidad Nacional de Tucumán, Ayacucho 471, T4000CAN San Miguel de Tucumán, Tucumán, República Argentina

A. Ben Altabef Facultad de Bioquímica, Química y Farmacia, Instituto de Química Física, INQUINOA-CONICET, Universidad Nacional de Tucumán, San Lorenzo 456, T4000CAN San Miguel de Tucumán, Tucumán, República Argentina

Chapter 1
Structural and Vibrational Analysis of Chromyl Chlorosulfate

Abstract In this chapter, a theoretical study of the structural and vibrational properties of the chromyl chlorosulfate compound using density functional theory (DFT) methods is presented. The results show two stable molecules for the chlorosulfate in gas phase and an average of both structures probably present in the solid phase. On the other hand, a complete assignment of all observed bands in the infrared spectrum for the compound was performed by combining DFT calculations with Pulay's Scaled quantum mechanics force field (SQMFF) methodology in order to fit the theoretical wavenumber values to the experimental ones. The calculations gave us precise knowledge of the normal modes of vibration taking into account the monodentate and bidentate coordination modes for the chlorosulfate ligands. The results were then used to predict the Raman spectra and molecular geometry of the compound, for which there are no experimental data. In this chapter, the scaled force constants and the scaling factors are also reported together with a comparison of the obtained values for similar compounds. Besides, the characteristics and nature of the Cr–O and Cr ← O bonds of the two stable structures were studied through the Wiberg's indexes calculated by means of the natural bond orbital (NBO) study, while the corresponding topological properties of the electronic charge density are analyzed by employing *Bader's* Atoms in the molecules theory (AIM).

Keyword · Chromyl chlorosulfate · Vibrational spectra · Molecular structure · Force field · DFT calculations

1.1 Introduction

The preparation of chromyl chlorosulfate, $CrO_2(SO_3Cl)_2$, was reported by Siddiqi et al. [1] by solvolytic reaction of CrO_2Cl_2 and CrO_3 with excess HSO_3Cl. From conductometric and spectrophotometric studies they found that the solutions of

S. A. Brandán, *A Structural and Vibrational Study of the Chromyl Chlorosulfate, Fluorosulfate, and Nitrate Compounds*, SpringerBriefs in Molecular Science, DOI: 10.1007/978-94-007-5763-9_1, © The Author(s) 2013

K_3CrO_4 and $K_2Cr_2O_7$ in chlorosulfuric acid [2] produce a similar highly conducting reddish-orange solution characteristic of the $CrO_2(SO_3Cl)_2$. In addition, they studied the magnetic susceptibility and the corresponding infrared spectrum of chromyl chlorosulfate in solid phase and only the main characteristics of this spectrum were published [1, 2]. Here, a detailed theoretical study of the structural and vibrational spectra of this compound is presented. In this case, an assignment of the observed bands was proposed by means of a normal coordinate analysis, considering the chlorosulfate groups as monodentate and bidentate ligands and accomplishing of a generalized valence force field (GVFF). For this purpose, the optimized geometry and frequencies for the normal modes of vibration were calculated at the B3LYP/6-31G* and B3P86/6-31G* theory levels. Then the performed calculations were used to predict the Raman spectrum for which no experimental data exists. The force field for the compound was obtained using transferable scaling factors for similar chromyl compounds [3–9]. In order to study the topological properties of electronic charge density and the nature of the two types of expected Cr–O and Cr ← O bonds, the Natural Bond Order (NBO) [10–13] and *Bader's* Atoms in Molecules theory (AIM) [14, 15] calculations were performed.

1.2 Geometry Study

In this compound, as in chromyl perchlorate [7, 8] and fluorosulfate compounds [9], two stable structures of C_2 symmetry were found, denominated $C_2(1)$ and $C_2(2)$. The numbering of the atoms for both structures is described in Figs. 1.1 and 1.2, respectively. For the SO_3Cl^- groups of the $C_2(1)$ structure both, monodentate and bidentate, coordination ligand types are expected. In the bidentate case, the two SO_3Cl^- groups are asymmetrically bonded to Cr, as in the experimental structure obtained for chromyl nitrate by Marsden et al. [16], while in the $C_2(2)$ structure the SO_3Cl^- groups can act only as monodentate ligands. Table 1.1 shows all calculations for the $C_2(2)$ structure resulting with positive frequency values, while for the $C_2(1)$ structure, imaginary frequency values by using all methods with the exception of B3PW91/6-31G*, B3P86/6-311G**, and B3P86/6-311++G** methods are observed [17–22]. The energy difference between both structures using B3P86/6-31G* calculations is low (1.84 kJ/mol), while with the B3LYP and B3PW91 methods, the size of the basis sets increases from 6-31G* to 6-311++G** and the dipole moment value decreases. On the other hand, the theoretical geometrical parameters of the chromyl group of chromyl chlorosulfate are compared in Table 1.2 with the experimental values of chromyl nitrate, because it is one of the few chromyl compounds whose structure is known. Accordingly, the method that best reproduces the experimental geometrical parameters for both structures of chromyl chlorosulfate is B3P86/6-31G*. For the $C_2(2)$ structure of this compound, a calculated low value of the O4–Cr1–O6 bond angle (between 106.5 and 109.1°), in relation to the experimental value of chromyl nitrate (140.5°), is observed.

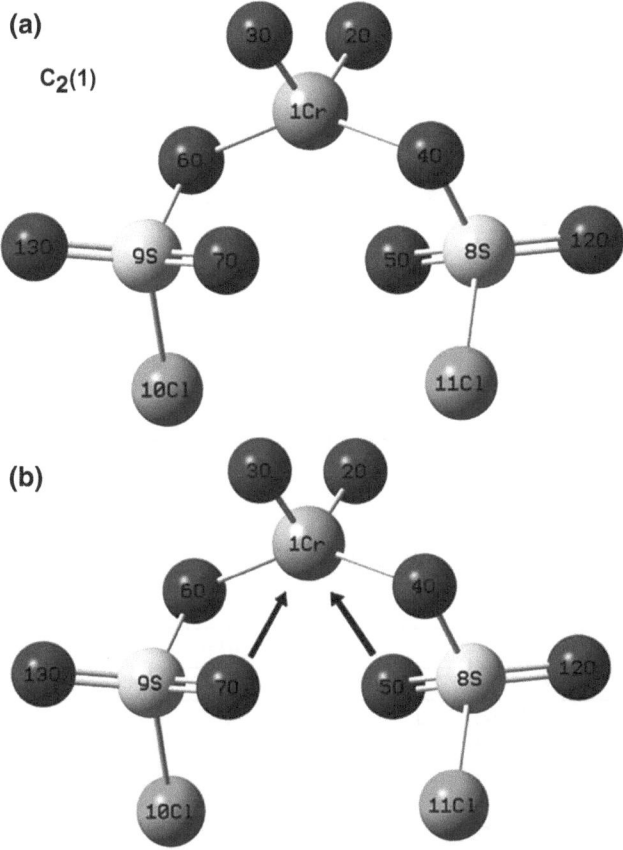

Fig. 1.1 The $C_2(1)$ molecular structure of chromyl chlorosulfate considering the chlorosulfate group as: **a** monodentate ligand and **b** bidentate ligand

A comparison of the bond orders, expressed by Wiberg's indexes, for both structures of chromyl chlorosulfate and fluorosulfate is shown in Table 1.3. In the $C_2(1)$ structures of these chromyl compounds the chromium atom forms six bonds, two Cr=O bonds, two Cr–O, and two Cr ← O; in the last case the bond order values of the Cl compound are high. These observations are justified because the F atom is more electronegative than the Cl atom and in this way the S–F bond is stronger than the S–Cl bond, while an inverse relation is observed in the bond order values of Cr=O bonds. On the contrary, the Cr atom of the $C_2(2)$ structures of both chromyl compounds forms only four bonds because the bond order values for the two Cr ← O bonds change in chromyl chlorosulfate at 0.0158. In chromyl chlorosulfate, as in chromyl nitrate [6] and perchlorate [7, 8], a contradiction with the Valence-Shell electron-pair repulsion (VSEPR) theory [23, 24] is observed due to which the DFT calculations predict that the O4–Cr1–O6 angles are higher than the O2=Cr1=O3 angle.

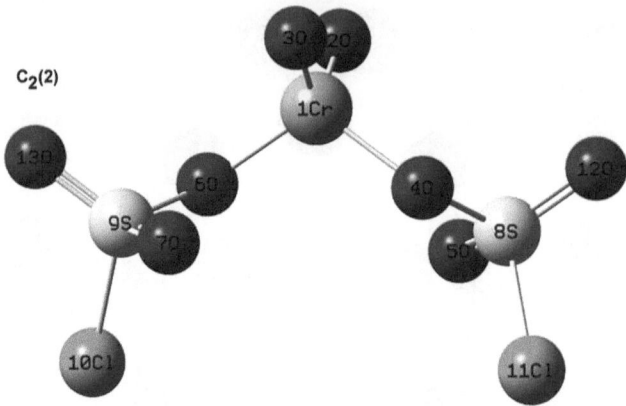

Fig. 1.2 The $C_2(2)$ molecular structure of chromyl chlorosulfate considering the chlorosulfate group as monodentate ligand

Table 1.1 Total energy (*ET*) and dipole moment (μ) for two structures of chromyl chlorosulfate using different theory levels

B3LYP method

$C_2(1)$ Symmetry			$C_2(2)$ Symmetry			ΔE
Basis set	*ET* (Hartree)	μ (D)	Basis set	*ET* (Hartree)	μ (D)	kJ/mol
6-31G*[a]	−3362.8647	0.25	6-31G*	−3362.8701	1.02	14.16
6-311G**[a]	−3363.1846	0.30	6-311G**	−3363.1980	2.28	35.15
6-311++G**[a]	−3363.2188	0.85	6-311++G**	−3363.2309	0.08	31.74
B3P86 Method						
6-31G*	−3365.7559	1.05	6-31G*	−3365.7566	1.10	1.84
6-311G**[a]	−3366.0627	0.70	6-311G**	−3366.0718	1.74	23.87
6-311++G**	−3366.0966	0.44	6-311++G**	−3366.1036	0.03	18.36
B3PW91 Method						
6-31G*	−3362.3805	0.89	6-31G*	−3362.3826	1.04	5.51
6-311G**[a]	−3362.6862	0.50	6-311G**	−3362.6981	2.29	31.48
6-311++G**[a]	−3362.7203	0.37	6-311++G**	−3362.7293	0.04	23.60

[a] Imaginary frequencies

This contradiction could be explained by analyzing the delocalized and/or bonding characters of the relevant molecular orbital [6–8, 25]. Here, the intermolecular interactions for the $C_2(1)$ structure have been studied using *Bader's* topological analysis of the charge electron density, and $\rho(r)$ by using the AIM program [15]. The analyses of the Cr–O and Cr $\leftarrow$ O bond critical points (BCPs) for the $C_2(1)$ structure with the B3P86/6-31G* method are reported and compared with the corresponding bidentate structure for chromyl fluorosulfate [9] by using a B3LYP/6-31G* level in Table 1.4.

Table 1.2 Comparison of experimental and calculated geometrical parameters at different theory levels for both structures of chromyl chlorosulfate

$CrO_2(SO_3Cl)_2$							$CrO_2(NO_3)_2$
Parameter	$C_2(1)$ Symmetry			$C_2(2)$ Symmetry			C_2 Symmetry
	B3LYP	B3PW91 6-31G*	B3P86	B3LYP	B3PW91 6-31G*	B3P86	Ref. [16]
Bond length (Å)							
R(1,2)	1.544	1.537	1.536	1.544	1.537	1.535	1.586 (2)
R(1,3)	1.544	1.537	1.536	1.544	1.537	1.535	1.586 (2)
R(1,4)	1.910	1.904	1.902	1.786	1.778	1.775	1.957 (5)
R(1,5)	2.360	2.327	2.307	3.269	3.247	3.223	2.254 (20)
R(1,6)	1.910	1.904	1.902	1.786	1.778	1.775	1.957 (2)
R(1,7)	2.360	2.327	2.307	3.269	3.247	3.223	2.254 (20)
R(4,8)	1.572	1.563	1.560	1.635	1.625	1.623	
R(5,8)	1.486	1.482	1.482	1.445	1.444	1.443	
R(6,9)	1.572	1.563	1.560	1.635	1.625	1.623	
R(7,9)	1.486	1.482	1.482	1.445	1.444	1.443	
R(8,11)	2.054	2.032	2.029	2.063	2.041	2.037	
R(8,12)	1.443	1.439	1.437	1.449	1.441	1.440	
R(9,10)	2.054	2.032	2.029	2.063	2.041	2.037	
R(9,13)	1.443	1.439	1.437	1.449	1.441	1.440	
RMSD	0.071	0.059	0.052	0.590	0.580	0.569	
Bond angle (°)							
A(2,1,3)	106.6	106.5	106.5	109.1	108.9	109.0	112.2 (71)
A(2,1,4)	103.8	103.7	103.7	110.5	110.4	110.3	97.2 (18)
A(2,1,5)	90.3	90.3	90.3	74.7	74.3	74.5	
A(2,1,6)	97.4	97.1	97.0	107.6	107.6	107.6	104.5 (9)
A(2,1,7)	159.3	159.5	159.6	154.0	153.8	154.7	
A(3,1,4)	97.4	97.1	97.0	107.6	107.6	107.6	104.5 (9)
A(3,1,5)	159.3	159.5	159.6	154.0	153.8	154.7	
A(3,1,6)	103.8	103.7	103.7	110.5	110.4	110.3	97.2 (18)
A(3,1,7)	90.3	90.3	90.3	74.7	74.3	74.5	
A(4,1,5)	66.4	66.9	67.2	49.5	49.7	50.2	
A(4,1,6)	144.0	144.8	144.9	111.2	111.6	111.7	140.5 (9)
A(4,1,7)	85.0	85.1	84.8	91.6	92.2	91.3	
A(5,1,6)	85.0	85.1	84.8	91.6	92.2	91.3	
A(5,1,7)	75.9	75.9	75.8	113.3	114.6	113.3	82.8 (60)
A(6,1,7)	66.4	66.9	67.2	49.5	49.7	50.2	
A(1,4,8)	103.6	102.9	102.5	126.5	126.5	125.7	
A(1,5,8)	88.0	88.2	88.5	68.6	68.8	69.1	
A(1,6,9)	103.6	102.9	102.5	126.5	126.5	125.7	
A(1,7,9)	88.0	88.2	88.5	68.6	68.8	69.1	
A(4,8,5)	101.5	101.5	101.4	108.7	108.7	108.6	
RMSD	6.3	6.4	6.5	17.6	17.8	17.4	

Table 1.3 Wiberg index bond matrix for both structures of chromyl chlorosulfate and fluorosulfate

Atoms	B3P86/6-31G*		B3LYP/6-31G*	
	$C_2(1)$ Symmetry		$C_2(2)$ Symmetry	
	[a]$CrO_2(SO_3Cl)_2$	[b]$CrO_2(SO_3F)_2$	[a]$CrO_2(SO_3Cl)_2$	[b]$CrO_2(SO_3F)_2$
1 Cr	0.0000	0.0000	0.0000	0.0000
2 O	2.0079	2.0077	1.9270	1.8991
3 O	2.0079	2.0077	1.9270	1.8991
4 O	0.4794	0.4844	0.6520	0.6909
5 O	0.1290	0.1196	0.0118	0.0158
6 O	0.4794	0.4844	0.6520	0.6909
7 O	0.1290	0.1196	0.0118	0.0158
8 S	0.0132	0.0135	0.0226	0.0077
9 S	0.0132	0.0135	0.0226	0.0077
10 X	0.0146	0.0078	0.0083	0.0073
11 X	0.0146	0.0078	0.0083	0.0073
12 O	0.0268	0.0294	0.0268	0.0282
13 O	0.0268	0.0294	0.0268	0.0282

X = F, Cl
[a] This work
[b] Ref. [9]

There are two important observations; first, the Cr1 ← O5 and Cr1 ← O7 BCP have the typical properties of the closed-shell interaction [6–9, 26] and the second observation is related to the topological properties of the Cr1–O4 and Cr1–O6 BCPs since in both cases they are the same. This calculated difference between the properties of the Cr–O and Cr ← O BCPs of chromyl chlorosulfate, in relation to fluorosulfate are justified because both results were performed with different

Table 1.4 Analysis of Cr ← O bond critical points in the $C_2(2)$ structure of chromyl chlorosulfate and fluorosulfate

$C_2(1)$ Structure of coordination bidentate				
Parameter[a]	[b]$CrO_2(SO_3Cl)_2$		[c]$CrO_2(SO_3F)_2$	
	B3P86/6-31G*		B3LYP/6-31G*	
	Cr1–O4/ Cr1–O6	Cr1 ← O5/ Cr1 ← O7	Cr1–O4/ Cr1–O6	Cr1 ← O5/ Cr1 ← O7
$\rho(r)$	0.1141	0.0411	0.1114	0.0359
$\nabla^2\rho(r)$	0.5367	0.1653	0.2090	0.1343
$\lambda 1$	−0.2148	−0.0452	−0.1900	−0.0374
$\lambda 2$	−0.1957	−0.0413	−0.1900	−0.0314
$\lambda 3$	0.9471	0.2518	0.9245	0.2031
$|\lambda 1|/\lambda 3$	0.2268	0.1795	0.2055	0.1841

[a] The quantities are in atomic units
[b] This work
[c] Ref. [9]

Table 1.5 Analysis of Cr–O bond critical points in the $C_2(2)$ structures of chromyl chlorosulfate, fluorosulfate, and perchlorate compounds at different theory levels

$C_2(2)$ Symmetry of coordination monodentate						
Parameter[a]	Cr1–O4/Cr1–O6					
	[b]$CrO_2(SO_3Cl)_2$		[c]$CrO_2(SO_3F)_2$		[d]$CrO_2(ClO_4)_2$	
	B3LYP 6-31G*	B3P86 6-31G*	B3LYP 6-31G*	B3P86 6-31G*	B3LYP 6-31G*	B3P86 6-31G*
$\rho(r)$	0.1471	0.1521	0.1478	0.1522	0.1532	0.1573
$\nabla^2\rho(r)$	0.7632	0.7852	0.7606	0.7915	0.7634	0.7787
$\lambda 1$	−0.2867	−0.2978	−0.2887	−0.2977	−0.3061	−0.3158
$\lambda 2$	−0.2638	−0.2754	−0.2673	−0.2762	−0.2785	−0.2874
$\lambda 3$	1.3138	1.3583	1.3167	1.3665	1.3480	1.3819
$|\lambda 1|/\lambda 3$	0.2182	0.2192	0.2192	0.2178	0.2271	0.2285

[a] The quantities are in atomic units
[b] This work
[c] Ref. [9]
[d] Refs. [7, 8]

methods. These two BCPs reveal that the coordination mode adopted for the chlorosulfate groups in the $C_2(1)$ structure is bidentate. On the other hand, for the $C_2(2)$ structure by using all calculations the Cr ← O BCPs were not observed, and for this, the coordination of the chlorosulfate groups in this structure is only possible as monodentate ligands.

The characteristic of these Cr–O BCPs with all methods used, compared with other chromyl compounds, can be seen in Table 1.5. Note that $\rho(r)$, by using B3LYP/6-31G* and B3P86/6-31G* methods, is slightly higher in chromyl perchlorate than the corresponding values of chromyl chlorosulfate and fluorosulfate, probably because the perchlorate groups have a major electronegativity of groups than the other ones.

1.3 Vibrational Study

Both structures of chromyl chlorosulfate have C_2 symmetry and 33 active vibrational normal modes in the infrared and Raman spectra (17 A + 16 B). Vibrational assignments were made taking into account both coordination modes by chlorosulfate groups on the basis of the PED in terms of symmetry coordinates and by comparison with similar molecules [1, 2, 6–9, 27, 28]. The observed frequencies and the assignment for chromyl chlorosulfate are given in Table 1.6. The theoretical infrared spectra for both structures are observed in Fig. 1.3, while the proposed theoretical Raman spectra are presented in Fig. 1.4. Some vibrational modes of different symmetries are calculated and mixed among them because the frequencies are approximately the same. The discussion of the assignment of the most important groups is presented below.

Table 1.6 Experimental and calculated frequencies (cm^{-1}), potential energy distribution, and assignment for chromyl chlorosulfate

Modes	ᵇTR	ᶜHSO₃Cl Raman	Assignment	ᵈRa	ᵉRa	B3P86 C₂(1) M	B3P86 C₂(1) B	B3LYP C₂(2) M	C₂(1) Monodentate	Bidentate	C₂(2) Monodentate
A Symmetry											
1	1170 s	1396 (3) dp	ν_a S–O₂	1195	1191	1408	1384	1416	ν_a S=O₂ ip	ν_s S=O	ν_a S=O₂ ip
2	1070 s	1153 (3) p	ν_s S–O₂	1050	1052	1126	1170	1184	ν_s S=O₂ op	ν_a S–O₂ op	ν_s S=O₂ ip
3	960 vs					997	1004	981	ν_s S–O₂	ν_s S–O₂ ip	ν_a Cr=O
4	960 vs	916 (3) p	ν S–OH			975	975	887	ν_s Cr=O	ν_s Cr=O	ν_s S–O₂
5	640 s	623 (3) p	δSO₂	563	567	653	686	653	ν_a Cr–O	δSO₂ op	ν_s Cr–O
6	580 s					615	631	603	δSO₂ op	ρSO₂ op	ρSO₂ op
7	570 m	513 (4)	ρSO₂	540	535	547	528	498	τwSO₂ ip	δSO₂ op	δSO₂ op
8	440 m	482 (1)	wag SO₂			462	486	444	δ CrO₂	wag SO₂ op	wag SO₂ op
9	440 m	416 (15) p	ν S–Cl			414	464	400	ν_s S–Cl	δCrO₂	δCrO₂
10	340 m			390	392	377	402	371	wag SO₂ op	δS–O–Cl ip	ν_s S–Cl
11	320 m	313 (7) dp	τwSO₂			306	317	283	τwSO₂ op	τwSO₂ ip	τwSO₂ op
12		298 (7)				239	256	209	ν_s Cr–O	νCr–O	ρ CrO₂
13		200 (2)	wag S–W	215	220	206	218	196	ρ CrO₂	ρCrO₂	δS–O–Cl ip
14						166	163	145	δ Cr–O–S ip	νCr–O	τw CrO₂ ip
15						132	141	87	τ Cr–O–S–O ip	τwCrO₂ ip	δa Cr–O–S
16						68	85	36	τw CrO₂ ip	δCrO₂ ipᶠ	τO–Cr–O–S op
17						45	60	17	τO–Cr–O–S ip	τCr–O–S–O ip	τO–Cr–O–S ip
B Symmetry											
18	1170 s	1396 (3) dp	ν_a S–O₂	1195	1191	1397	1370	1416	ν_a S=O₂ op	ν_a S=O	ν_a S=O₂ op
19	1070 s	1153 (3) p	ν_s S–O₂	1050	1052	1138	1180	1187	ν_s S=O₂ ip	ν_a S–O₂ ip	ν_s S=O₂ op
20	960 vs					980	982	972	ν_a S–O₂	ν_s S–O₂ op	ν_s Cr=O

(continued)

Table 1.6 (continued)

Modes	bIR (Exp)	Raman (cHSO$_3$Cl)	dRa	eRa	Assignment (cHSO$_3$Cl)	B3P86 C$_2$(1) M	B3P86 C$_2$(1) B	B3LYP C$_2$(2) M	aAssignment C$_2$(1) Monodentate	aAssignment C$_2$(1) Bidentate	aAssignment C$_2$(2) Monodentate
21	960 vs	916 (3) p			ν S–OH	952	953	804	ν_a Cr=O	ν_a Cr=O	ν_a S–O$_2$
22	640 s	623 (3) p			δSO$_2$	671	715	684	ρSO$_2$ op	δSO$_2$ ip	ν_a Cr–O
23	580 s		563	567		628	643	605	ν_a S–Cl	ν_a S–Cl	τwSO$_2$ ip
24	570 m	513 (4)	540	535	ρSO$_2$	550	540	503	δSO$_2$ ip	ρSO$_2$ ip	δSO$_2$ ip
25	440 m	482 (1)			wag SO$_2$	415	483	417	ρ SO$_2$ ip	wag SO$_2$ ip	ρSO$_2$ ip
26	440 m	416 (15) p			ν S–Cl	396	409	380	ρ SO$_2$ ip	δS–O–Cl op	ν_a S–Cl
27	340 m		390	392		318	338	331	wag CrO$_2$	δSO$_2$ op	wag CrO$_2$
28	320 m	313 (7) dp			τwSO$_2$	292	307	274	wag SO$_2$ ip	τwSO$_2$ op	wag SO$_2$ ip
29		298 (7)				214	224	202	δS–O–Cl op	wag CrO$_2$	τw CrO$_2$ op
30		200 (2)	215	220	wag S–W	187	207	174	τw CrO$_2$ op	τwCrO$_2$ op	δS–O–Cl op
31						118	126	83	δ Cr–O–S op	τCr–O–S–O op	δs Cr–O–S
32						66	80	18	τCr–O–S–O op	δCrO$_2$ op[f]	τ Cr–O–S–O op
33						41	43	13	τO–Cr–O–S op	ν Cr–O	τCr–O–S–O ip

Abbreviations: ν stretching, δ deformation, ρ rocking, wag wagging, τw torsion, a antisymmetric, s symmetric, op out of phase, ip in phase, M monodentate, B bidentate, v very, s strong, m medium, w weak, sh shoulder, p polarized, dp depolarized

a This work
b Ref. [1, 2] CrO$_2$(SO$_3$Cl)$_2$
c Refs. [27, 28] HSO$_3$Cl
d Ref. [21] NaSO$_3$Cl in dimethyl sulfoxide
e Ref. [28] KSO$_3$Cl in dimethyl sulfoxide
f O–Cr–O deformation

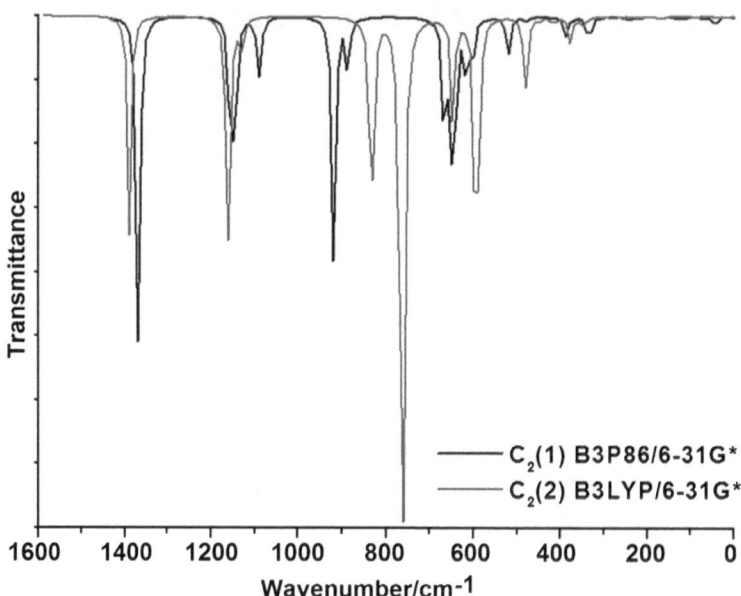

Fig. 1.3 Theoretical infrared spectra of chromyl chlorosulfate for the C$_2$(1) structure at B3P86/6–31G* level and for the C$_2$(2) structure at B3LYP/6–31G* level

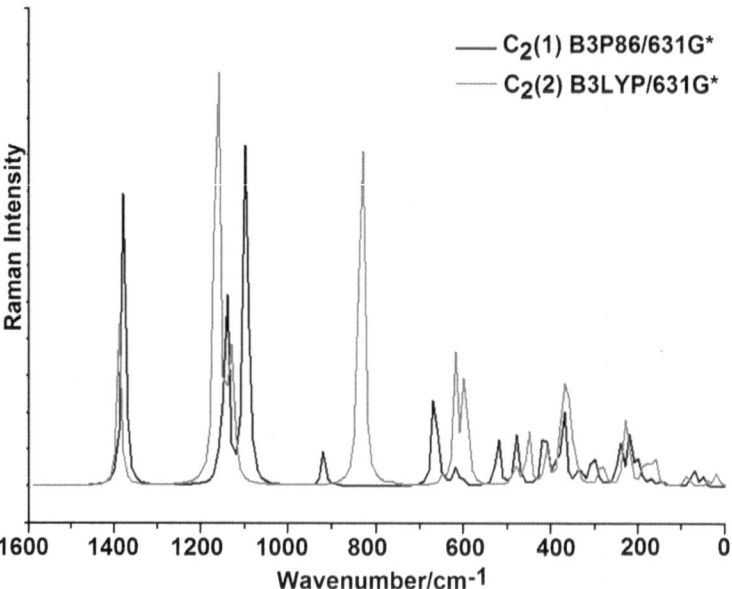

Fig. 1.4 Raman spectra for the theoretical C$_2$(1) and C$_2$(2) structures of chromyl chlorosulfate at B3P86/6–31G* and B3LYP/6–31G* theory levels, respectively

1.4 Coordination Monodentate of the Chlorosulfate Groups

1.4.1 $C_2(1)$ and $C_2(2)$ Structures

The IR frequencies and the PED obtained by B3P86/6-31G* and B3LYP/6-31G* calculations for both structures with monodentate coordination appear in Tables 1.7 and 1.8. The covalent bonding of the chlorosulfate group is easily recognized from its infrared spectrum because the symmetry group changes from the point group C_{3V} of the free ion to C_2 of the compound.

1.4.2 Chlorosulfate Groups

In the HSO_3Cl compound [27, 28] the two $S=O_2$ antisymmetric and symmetric stretching modes were assigned at 1396 and 1153 cm^{-1}. In $CrO_2(SO_3Cl)_2$, these modes were previously assigned by Siddiqi et al. at 1170 and 1070 cm^{-1}, respectively [1, 2]. Here, for both structures, the $S=O_2$ in-phase and out-of-phase antisymmetric stretching modes and the corresponding symmetric stretching modes are split by less than 14 cm^{-1}, while the antisymmetric and symmetric modes are split by more than 214 cm^{-1}, indicating in this last case a strong contribution of the Cr central atom in these vibrations. In the calculated infrared spectrum of the $C_2(1)$ structure (Fig. 1.3) the bands associated with these modes are observed with different intensities at 1369 and 1154 cm^{-1}, while in the corresponding spectrum of the $C_2(2)$ structure the bands are observed with the same intensities at 1391 and 1159 cm^{-1}. In the Raman spectra of both structures the bands are observed with inverted intensities (Fig. 1.4). Experimentally, in the chlorosulfate compounds the S–Cl stretching modes are observed between 540 and 416 cm^{-1} [27, 28]. In $CrO_2(SO_3Cl)_2$, the stretching mode was assigned by Siddiqi et al. [1, 2] to the IR band at 440 cm^{-1}. Here, according to calculations for the $C_2(1)$ structure, the S–Cl antisymmetric and symmetric modes are assigned at 580 and 440 cm^{-1}, respectively, while for the $C_2(1)$ structure the two modes are assigned to the IR bands at 440 and 340 cm^{-1}. The SO_2 antisymmetric and symmetric deformation modes were assigned by Siddiqi et al. [1, 2] to the bands at 580 and 570 cm^{-1}, respectively. In the $C_2(2)$ structure, the band at 580 cm^{-1} is associated with the SO_2 out-of-phase deformation mode while the band at 570 cm^{-1} is associated with both deformation modes of the $C_2(2)$ structure and with the in-phase deformation mode of the $C_2(1)$ structure. The SO_2 in-phase and out-of-phase rocking modes were not previously assigned by Siddiqi et al. [1, 2].

In the $C_2(1)$ structure, the bands at 640 and 440 (PED 47 %) cm^{-1} are clearly assigned to the SO_2 out-of-phase and in-phase rocking modes, respectively, while in the $C_2(2)$ structure these modes are associated with the bands at 580 and 440 cm^{-1}. The SO_2 wagging and twisting modes are observed in the low frequencies region [27, 28], thus, in the $C_2(1)$ structure, the SO_2 out-of-phase and

Table 1.7 Experimental and calculated frequencies (cm^{-1}), potential energy distribution, and assignment for the $C_2(1)$ Structure of chromyl chlorosulfate

$^a C_2(1)$ Structure of monodentate coordination

Modes	Observed[b]	Calculated[c]	SQM[d]	PED (≥10 %)
A Symmetry				
1	1170 s	1379	1408	ν_a SO$_2$ ip (78)
2	1070 s	1141	1126	ν_s SO$_2$ ip (70) + ν_a SO$_2$ ip (16)
3	960 vs	1099	997	ν_s SO$_2$ (74)
4	960 vs	920	975	ν_s Cr=O (95)
5	640 s	646	653	ν_a Cr–O(35) + τSO$_2$ ip (27) + ρ SO$_2$ ip (15) + δ Cr–O–S op (11)
6	580 s	604	615	τSO$_2$ ip(27) + ρSO$_2$ op(16) + wag SO$_2$ op(12) + ν_s S–Cl (11) + ν_s Cr–O(11)
7	570 m	521	547	τSO$_2$ ip(45) + δ SO$_2$ op(36)
8	440 m	478	462	δ CrO$_2$ (61) + τSO$_2$ ip(10)
9	440 m	414	414	τSO$_2$ ip(29) + ν_s S–F(20) + ρSO$_2$ op(16) + wag SO$_2$ op (13)
10	340 m	372	377	wag SO$_2$ op (32) + δSO$_2$ ip (11)
11	320 m	309	306	τSO$_2$ op (29) + δ O–S–Cl ip (20) + τ O–Cr–O–S ip (12) + δ SO$_2$ ip(10) + wag SO$_2$ op (10) +τSO$_2$ ip(10)
12		243	239	τSO$_2$ op (23) + ν_s Cr–O (20) + δ O–S–Cl ip (17) + ρSO$_2$ op(12)
13		217	206	ρ CrO$_2$ (56)
14		171	166	δ Cr–O–S ip (42) + ν_s Cr–O (13) + τw CrO$_2$ ip (11) + wag SO$_2$ op (10) + τSO$_2$ ip (10)
15		146	132	τ Cr–O–S–O ip (35) + τw CrO$_2$ ip (23) + ρ CrO$_2$ (16)
16		73	68	τ O–Cr–O–S ip (59) + τw CrO$_2$ ip (16)
17		50	45	τ Cr–O–S–O ip (54) + τ O–Cr–O–S ip (38)
B Symmetry				
18	1170 s	1367	1397	ν_a SO$_2$ op (78)
19	1070 s	1154	1138	ν_s SO$_2$ ip (69) + ν_a SO$_2$ op (17)
20	960 vs	1088	980	ν_a SO$_2$ (46) + ν_a Cr=O (37)
21	960 vs	888	952	ν_a Cr=O (53) + ν_a SO$_2$ (42)
22	640 s	666	671	ρ SO$_2$ op (27) + ν_s Cr–O (22) + δ Cr–O–S ip (17) + wag SO$_2$ op (15)

(continued)

Table 1.7 (continued)

[a]$C_2(1)$ Structure of monodentate coordination

Modes	Observed[b]	Calculated[c]	SQM[d]	PED ($\geq$10 %)
23	580 s	619	628	ν_a S–Cl (31) + δSO$_2$ op (17) + ν_a Cr–O (15) + τSO$_2$ ip (11)
24	570 m	523	550	ρ SO$_2$ op (38) + δSO$_2$ ip (31) + τSO$_2$ ip (17)
25	440 m	415	415	ρ SO$_2$ ip (47) + ν_a S–Cl (19)
26	440 m	389	396	ρ SO$_2$ ip (25) + ν_a S–Cl (22) + τSO$_2$ ip(20) + δSO$_2$op (15) + ν_a Cr–O(12)
27	340 m	335	318	wag CrO$_2$ (30) + δ O–S–Cl op (14) + τSO$_2$ ip (17)
28	320 m	297	292	wag SO$_2$ ip (50) + ρ SO$_2$ ip (15) + τSO$_2$ ip (11)
29		225	214	wag CrO$_2$ (44) + δ O–S–Cl op (29)
30		197	187	τw CrO$_2$ op (29) + τ O–Cr–O–S op (21) + δ O–S–Cl op (12) + wag SO$_2$ ip (11)
31		124	118	τw CrO$_2$ op (27) + δ Cr–O–S op (24)
32		73	66	τ Cr–O–S–O op (76)
33		44	41	δ Cr–O–S op (24) + τ O–Cr–O–S op (22) + ρ SO$_2$ ip (13)

[a] This work
[b] From Refs. [1, 2]
[c] DFT B3P86/6-31G*
[d] From scaled quantum mechanics force field

Table 1.8 Experimental and calculated frequencies (cm^{-1}), potential energy distribution, and assignment for the $C_2(2)$ structure of chromyl chlorosulfate

[a]$C_2(2)$ structure of monodentate coordination

Modes	Observed[b]	Calculated[c]	SQM[d]	PED ($\geq 10\ \%$)
A Symmetry				
1	1170 s	1389	1416	ν_a SO_2 ip (89)
2	1070 s	1161	1184	ν_s SO_2 ip (93)
3	960 vs	1143	981	ν_s Cr=O (97)
4	960 vs	834	887	ν_s SO_2 (55) + ν_s Cr–O(30)
5	640 s	618	653	ν_s Cr–O(29) + δSO_2 ip (21) + ν_a SO_2 op (19) + δ O–S–Cl ip (10)
6	580 s	589	603	ρSO_2 op(50) + ν_s S–Cl (17) + τSO_2 ip(17)
7	570 m	482	498	δ SO_2 op(45) + ρSO_2 ip(25)
8	440 m	453	444	wag SO_2 op (48) + τSO_2 ip(19) + δ CrO_2 (19)
9	440 m	409	400	δ CrO_2 (36) + wag SO_2 op (21) + τw CrO_2 ip (16)
10	340 m	364	371	ν_s S–Cl (28) + τSO_2 ip(22) + ρSO_2 op(18)
11	320 m	288	283	τSO_2 op (40) + wag SO_2 op (33) +τSO_2 ip(12)
12		230	209	ρ CrO_2 (81)
13		184	196	δ O–S–Cl ip (10) + ν_s Cr–O (10)
14		159	145	τw CrO_2 ip (68)
15		92	87	δ O–S–Cl op (67) + τw CrO_2 *o*p (10) + τ Cr–O–S–O op (10)
16		40	36	τ O–Cr–O–S op (55) + τ Cr–O–S–O op (26) + δ O–S–Cl ip (10)
17		19	17	τ O–Cr–O–S ip (74) + τ Cr–O–S–O ip (10)
B Symmetry				
18	1170 s	1389	1416	ν_a SO_2 op (89)
19	1070 s	1163	1187	ν_s SO_2 op (92)
20	960 vs	1128	972	ν_s Cr=O (95)
21	960 vs	763	804	ν_a SO_2 (56) + ν_a Cr–O (41)
22	640 s	648	684	ν_a Cr–O (30) + ν_a SO_2 (27) + δ O–S–Cl op (12) + τSO_2 ip(10)
23	580 s	596	605	τSO_2 ip (38) + ν_a S–Cl (27) + δSO_2 op (18)
24	570 m	484	503	τSO_2 ip (31) + δSO_2 ip (25) + ρ SO_2 op (20) + wag SO_2 ip (10)
25	440 m	415	417	ρ SO_2 ip (49) + τSO_2 ip (42)
26	440 m	378	380	ν_a S–Cl (77)
27	340 m	346	331	wag CrO_2 (45) +τSO_2 ip (22) + δ O–S–Cl op (17)
28	320 m	278	274	wag SO_2 ip (55) + ρ SO_2 ip (29) + τSO_2 ip(12)
29		222	202	τw CrO_2 op (68) + δ Cr–O–S op (10)
30		169	174	δ O–S–Cl op (45) + wag CrO_2 (28) + ν_a Cr–O (11)
31		86	83	δ Cr–O–S ip (78)
32		20	18	τ Cr–O–S–O op (64) + δ Cr–O–S op (15)
33		15	13	τ O–Cr–O–S ip (75)

[a] This work
[b] From Refs. [1, 2]
[c] DFT B3LYP/6-31G*
[d] From scaled quantum mechanics force field

in-phase waggings are clearly assigned at 340 and 320 cm^{-1}, respectively, while in the $C_2(2)$ structure the bands located at 440 and 320 cm^{-1} are associated with those vibrational modes. Previously, the S–Cl wagging mode was associated by Siddiqi et al. [1, 2] with the band at 320 cm^{-1} and the SO_2 in-phase and out-of-phase torsion modes were not assigned. Here, it is possible to assign these modes in the $C_2(1)$ structure at 570 (PED 45 %) and 320 cm^{-1} and in the $C_2(2)$ structure these modes are assigned to the bands at 580 and 320 cm^{-1}, respectively. The in-phase and out-of-phase S–O–Cl deformation modes for both structures were not assigned because they are predicted in the 166–83 cm^{-1} region (Table 1.6).

1.4.3 Chromyl Group

In the chromyl compounds the Cr=O antisymmetric and symmetric stretchings appear in the 1050–900 cm^{-1} region [6–9]. In the $C_2(1)$ structure, these modes are split by more than 20 cm^{-1}, while in the $C_2(2)$ structure the split is of 9 cm^{-1}, indicating a low contribution of the Cr central atom in these vibrations. In both structures these modes are calculated in the expected region, hence, they are assigned to the strong IR band at 960 cm^{-1} as reported by Siddiqi et al. [1, 2]. In the Infrared spectra of the two structures the bands associated with both Cr=O stretching modes are observed with inverted intensities (Fig. 2.3). The CrO_2 bending mode was not previously assigned, in this case it is assigned for both structures, as the theoretical calculations predict, at 440 cm^{-1}. The calculations predict the wagging, rocking, and twisting modes of the CrO_2 group in the low frequencies region and coupled with other modes of the chlorosulfate groups. The assignment of those modes for both structures are very different among themselves, as observed in Table 1.6, with the exception of the wagging modes, which in both structures are assigned to the band at 340 cm^{-1}. Thus, in the $C_2(1)$ structure the rocking and twisting modes are calculated, respectively, at 206, 187, and 68 cm^{-1}, while in the $C_2(2)$ structure the modes are calculated at 209, 202, and 145 cm^{-1}.

1.5 Coordination Bidentate of the Chlorosulfate Groups

1.5.1 $C_2(1)$ Structure

The observed and calculated IR frequencies and PED obtained by B3P86/6-31G* calculations considering the bidentate coordination appear in Table 1.9.

Table 1.9 Experimental and calculated frequencies (cm^{-1}), potential energy distribution, and assignment for the $C_2(1)$ structure of chromyl chlorosulfate

[a]$C_2(1)$ Structure of bidentate coordination

Modes	Observed[b]	Calculated[c]	SQM[d]	PED (≥ 10 %)
A Symmetry				
1	1170 s	1379	1384	ν_s S=O (52) + ν_s SO$_2$ ip (31) + ν_a SO$_2$ op (10)
2	1070 s	1141	1170	ν_a SO$_2$ op (51) + ν_s S=O (38)
3	960 vs	1099	1004	ν_s SO$_2$ ip (56) + ν_a SO$_2$ ip (32)
4	960 vs	920	975	ν_s Cr=O (95)
5	640 s	647	686	δ SO$_2$ op (54) + ν_a S–Cl (18) + ν_a Cr–O (11)
6	580 s	628	631	ν_s S–Cl (35) + δ SO$_2$ ip (22) + wag SO$_2$ op (17) + ν_sCr–O (15)
7	570 m	526	528	ρ SO$_2$ op (44) + δ SO$_2$ ip (14) + ν Cr–O (12) + wag SO$_2$ op (10)
8	440 m	481	486	wag SO$_2$ op (37) + ν_s S–Cl (33) + δ O–S–Cl ip (17)
9	440 m	478	464	δCrO$_2$ (63) + τw CrO$_2$ ip (14)
10	340 m	391	402	δO–S–Cl ip (42) + δCrO$_2$ (12)
11	320 m	320	317	τSO$_2$ ip (59) δ O–S–Cl ip (11)
12		253	256	δO–S–Cl ip(23) +ν_sCr–O (19) +τ CrO$_2$ ip(18) + δSO$_2$ ip (15) + ρ SO$_2$op (11)
13		231	218	ρ CrO$_2$ (65)
14		172	163	ν Cr–O (43) + ν_s Cr–O (32)
15		147	141	τw CrO$_2$ ip (41) + δCrO$_2$ ip (29) + τ Cr–O–S–O ip (19)
16		89	85	δ Cr–O$_2$ ip (29) + ρ CrO$_2$ (27) + τSO$_2$ ip (20) + τw CrO$_2$ ip (11)
17		61	60	τ Cr–O–S–O ip (75) + δ Cr–O$_2$ ip (15)
B Symmetry				
18	1170 s	1368	1370	ν_a S=O (53) + ν_s SO$_2$ ip (30) + ν_a SO$_2$ ip (11)
19	1070 s	1154	1180	ν_a SO$_2$ ip (47) + ν_a S=O (40)
20	960 vs	1088	982	ν_s SO$_2$ op (42) + ν_a Cr=O (34) + ν_a SO$_2$ ip (17)
21	960 vs	888	953	ν_a Cr = O (49) + ν_a SO$_2$ ip (24) +ν_s SO$_2$ op (22)
22	640 s	673	715	δSO$_2$ ip (56) + ν_s Cr–O (16) + ν_s S–Cl (15)
23	580 s	645	643	ν_a S–Cl (31) + δSO$_2$ op (20) + ν_a Cr–O (17) + wag SO$_2$ ip (16)
24	570 m	526	540	ρ SO$_2$ ip (32) + δSO$_2$ op (23) + ν Cr–O (17)
25	440 m	479	483	wag SO$_2$ ip (40) + ν_a S–Cl (32) + δ O–S–Cl op (15)
26	440 m	407	409	δO–S–Cl op (42) + ρ SO$_2$ ip (18) + ν_a Cr–O (15) + wag SO$_2$ ip (14)
27	340 m	342	338	wag CrO$_2$ (32) + δ O–S–Cl op (32) + δSO$_2$ op (10)
28	320 m	309	307	τSO$_2$ op (50) + wag CrO$_2$ (10)
29		234	224	wag CrO$_2$ (45) ρ SO$_2$ ip (13) +τw CrO$_2$ op (13) + τSO$_2$ op (11)
30		215	207	τw CrO$_2$ op (56) + δCrO$_2$ op (11)
31		131	126	ν Cr–O (32) + ν_a Cr–O (26) + τ Cr–O–S–O op (10)
32		81	80	τ Cr–O–S–O ip (67) + δCrO$_2$ op (18)
33		45	43	ν Cr–O (37) + ν_a Cr–O (28) + τw CrO$_2$ op (17)

[a] This work

[b] From Refs. [1, 2]

[c] DFT B3P86/6-31G*

[d] From scaled quantum mechanics force field

1.5.2 Chlorosulfate Groups

In this case the S=O symmetric stretching mode is calculated at 1384 cm^{-1}, while the corresponding antisymmetric mode at 1370 cm^{-1}, for this, the strong IR band at 1170 cm^{-1} is assigned to these modes. The assignment of the bands associated with the SO$_2$ out-of-phase and in-phase antisymmetric and symmetric modes is similar to the monodentate type, as can be seen in Table 1.6, while the SO$_2$ in-phase and out-of-phase deformation modes are assigned to the band at 640 cm^{-1}. The SO$_2$ out-of-phase and in-phase wagging modes are assigned to the band at 440 cm^{-1} while both rocking modes are calculated at 540 and 528 cm^{-1} for this, those modes are assigned at 570 cm^{-1}.

Finally, the band at 320 cm^{-1} is associated with both SO$_2$ out-of-phase and in-phase twisting modes, because in chromyl fluorosulfate these modes were calculated in this region [9].

1.5.3 Chromyl Group

For a bidentate coordination of the chlorosulfate groups, two Cr=O stretching modes and four Cr–O stretching modes are expected. The Cr=O stretchings are clearly assigned at 960 cm^{-1}, because normally both modes are observed in the 1050–900 cm^{-1} region [1, 2, 6–9]. Here, the CrO$_2$ bending (O=Cr=O) mode appears at the same wavenumbers as in the monodentate case of the C$_2$(2) structure. Hence, this mode is assigned to the band at 440 cm^{-1}. Only three of the four expected Cr–O stretching modes are calculated with higher PED contribution at 256, 163, and 43 cm^{-1} and, as a consequence, these bands are associated with those stretching modes. The two modes, CrO$_2$ in-phase and out-of-phase bending modes (O–Cr–O) are calculated at lower wavenumbers than the chromyl fluoro-sulfate (128 and 148 cm^{-1}) [9] and for this, both modes calculated at 85 and 80 cm^{-1}, respectively, were not assigned. The wagging, rocking, and twisting modes of the CrO$_2$ group are calculated strongly mixed with other modes. The rocking and the out-of-phase twisting mode are assigned as in the monodentate case of the C$_2$(1) structure (Table 1.8) while the wagging and in-phase twisting mode are calculated and assigned, respectively, at 224 and 141 cm^{-1}.

1.6 Force Field

The harmonic force field for chromyl chlorosulfate and the force constants were calculated using the SQM procedure [29] with the MOLVIB program [30, 31]. A comparison of the force constants for both structures of chromyl chlorosulfate with those corresponding to chromyl fluorosulfate [9] and nitrate [6] appears in Table 1.10.

Table 1.10 Comparison of scaled internal force constants for chromyl chlorosulfate with similar compounds

| Force constant | B3P86/6-31G* [a]CrO$_2$(SO$_3$Cl)$_2$ C$_2$(1) | | B3LYP/6-31G* [b]CrO$_2$(SO$_3$F)$_2$ | | | | [c]CrO$_2$(NO$_3$)$_2$ | |
| | | | C$_2$(2) | C$_2$(1) | | C$_2$(2) | C$_2$ | |
	M	B	M	M	B	M	M	B
f (S=O)	10.1	10.0	10.6	10.1	10.0	10.6		
f (S–O)	6.6	8.6	4.4	6.4	8.6	4.7		
f (Cr=O)	6.8	6.8	6.7	6.6	6.6	8.8	6.55	6.53
f (Cr–O)	2.8	1.4	3.8	2.8	1.3	3.7	6.09	1.44
f (S–X)	2.7	2.7	2.4	5.2	5.2	4.8		
f (O=S=O)	2.2	4.9	2.0	2.2	4.8	1.6		
f (O=Cr=O)	2.6	2.4	1.9	2.4	2.3	2.1	2.53	1.66
f (O–Cr–O)		0.9			0.8		0.80	0.93
f (S–O–Cr)	2.1		0.4	1.8		0.4		
f (O–S–X)	1.8	2.4	1.7	2.0	2.6	1.7		

Units: mdyn Å^{-1} for stretching and stretching–stretching interaction and mdyn Å rad^{-2} for angle deformations. Abbreviations: M monodentate, B bidentate. X = F, Cl
[a] This work
[b] Ref. [9]
[c] Ref. [6]

　　　The calculated f(S=O), f(Cr=O), f(S–X), and f(O=Cr=O) force constants for the bidentate and monodentate coordination modes of the chlorosulfate groups in the C$_2$(1) structure are approximately the same, while the f(S–O), f(Cr–O), f(O=S=O), and f(O–S–X) force constants change with the coordination mode. Thus, the higher value in the f(S–O) force constant in the bidentate coordination (8.6 mdyn Å^{-1}), in relation to the monodentate coordination (6.4 mdyn Å^{-1}), is justified because there are two S–O stretchings in each chlorosulfate group. Whereas the higher value in the f(Cr–O) force constant in the monodentate coordination (2.8 mdyn Å^{-1}), as were also observed in CrO$_2$(NO$_3$)$_2$ [6] and CrO$_2$(SO$_3$F)$_2$ [9], it is justified because in the bidentate coordination there are four Cr–O strechings. Also, these reasons justify the higher value of the f(O=S=O) force constant in a bidentate coordination. On the other hand, the differences between the geometrical parameters of both monodentate structures justify the force constant values. Thus, the lower values of the force constants of O=S=O deformations in the C$_2$(2) structure of chromyl chlorosulfate and fluorosulfate [9] can be attributed to the higher values of the O=S=O angles (122.9° in chromyl chlorosulfate) in this structure than in the other ones (120.8° in chromyl chlorosulfate). In similar form, the lower f(O–S–X) force constant values in the C$_2$(2) structure of both chromyl compounds are explained. The S=O stretching force constants considering the two structures and coordination modes are near to the values reported for chromyl fluorosulfate [9] and are independent of the method used. By contrast, the analysis of the force constants of chromyl chlorosulfate with the values for CrO$_2$(NO$_3$)$_2$ suggests that in the monodentate coordination the nature of the anion linked to the chromyl group has influence on the f(Cr–O) force constant value; thus, this value is higher in $CrO_2(NO_3)_2$ (6.09 mdyn $Å^{-1}$) [6] than chromyl perchlorate (2.54 mdyn $Å^{-1}$) [7, 8] and fluorosulfate (2.8 mdyn $Å^{-1}$) [9], and moreover, is strongly dependent on the method used.

1.7 Conclusions

In this chapter the calculations suggest the existence of two molecular $C_2(1)$ and $C_2(2)$ structures for chromyl chlorosulfate, both of C_2 symmetry, which probably origin the different colorations observed in different preparations of the compound.

The B3P86/6-31G* method for $C_2(1)$ structure and B3LYP/6-31G* level for $C_2(2)$ structure were employed to obtain a molecular force field and vibrational frequencies.

The presence of both coordination modes was detected in the IR spectrum, and a complete assignment of the 33 normal vibration modes corresponding to chromyl chlorosulfate are reported.

The NBO and AIM studies confirm the hexacoordination of the Cr atom for the $C_2(1)$ structure of chromyl chlorosulfate.

Acknowledgments This work was subsidized with grants from CIUNT (Consejo de Investigaciones, Universidad Nacional de Tucumán). The author thanks Prof. Tom Sundius for his permission to use MOLVIB.

References

1. Z.A. Siddiqi, Lutfullah, S.A.A. Zaidi, Bull. Soc. Chim. Fr. **11–12**, 466 (1980)
2. Z.A.Siddiqi, Lutfullah, N.A. Ansari, S.A. Zaidi, J Inorg. Nucl. Chem. **43**, 397 (1981)
3. G. Rauhut, P. Pulay, J. Phys. Chem. **99**, 3093 (1995)
4. G. Rauhut, P. Pulay, J. Phys. Chem. **99**, 14572 (1995)
5. F. Kalincsák, G. Pongor, Spectrochim. Acta A **58**, 999 (2002)
6. S.A. Brandán, M.L. Roldán, C. Socolsky, A. Ben Altabef, Spectrochim. Acta, Part A **69**, 1027 (2008)
7. S.A. Brandán, J. Mol. Struc. Theochem. **908**, 19 (2009)
8. S.A. Brandán, in *Structural and Vibrational Properties of Chromyl Perchlorate*, ed. by L.E. Mattews, Chapter 3 (Nova Science Publisher, Inc., Hauppauge, New York, 2010)
9. A. Ben Altabef, S.A. Brandán, J. Mol. Struc. **981**, 146 (2010)
10. A.E. Reed, L.A. Curtis, F. Weinhold, Chem. Rev. **88**(6), 899 (1988)
11. J.P. Foster, F. Weinhold, J. Am. Chem. Soc. **102**, 7211 (1980)
12. A.E. Reed, F. Weinhold, J. Chem. Phys. **83**, 1736 (1985)
13. E.D. Glendening, J.K. Badenhoop, A.D. Reed, J.E. Carpenter, F. Weinhold, NBO 3.1 (Theoretical Chemistry Institute, University of Wisconsin; Madison, WI 1996)
14. R.F.W. Bader, Atoms in molecules, a quantum theory (Oxford University Press, Oxford, ISBN: 0198558651, 1990)

15. F. Biegler-Köning, J. Schönbohm, D. Bayles, AIM2000: a program to analyze and visualize atoms in molecules. J. Comput. Chem. **22**, 545 (2001)
16. C.J. Marsden, K. Hedberg, M.M. Ludwig, G.L. Gard, Inorg. Chem. **30**, 4761 (1991)
17. A.B. Nielsen, A.J. Holder, *GaussView,* User's Reference (Gaussian Inc., Pittsburgh, PA, USA, 2000–2003)
18. Gaussian 03, Revision B.01, M.J. Frisch, J.A. Pople, Gaussian, Inc., Pittsburgh PA, (2003)
19. A.D. Becke, J. Chem. Phys. **98**, 5648 (1993)
20. C. Lee, W. Yang, R.G. Parr, Phys. Rev. B **37**, 785 (1988)
21. P.J. Perdew, Phys. Rev. B **33**, 8822 (1986)
22. P.J. Perdew, K. Burke, Y. Wang, Phys. Rev. B **54**, 16533 (1996)
23. R.J. Gillespie (ed.), *Molecular Geometry* (Van Nostrand-Reinhold, London, 1972)
24. R.J. Gillespie, I. Bytheway, T.H. Tang, R.F.W. Bader, Inorg. Chem. **35**, 3954 (1996)
25. M. Fernández Gómez, A. Navarro, S.A. Brandán, C. Socolsky, A. Ben Altabef, E.L. Varetti, J. Mol. Struct. (Theochem) **626**, 101 (2003)
26. G.L. Sosa, N. Peruchena, R.H. Contreras, E.A. Castro, J. Mol. Struct. (Theochem) **401**, 77 (1997)
27. A. Simon, H. Kriegsmann, H. Dutz, Ber. **89**, 2378 (1956)
28. R.J. Gillespie, E.A. Robinson, Can. J. Chem. **40**, 644 (1962)
29. P. Pulay, G. Fogarasi, F. Pang, J.E. Boggs, J. Am. Chem. Soc. **101**, 2550 (1979)
30. T. Sundius, J. Mol. Struct. **218**, 321 (1990)
31. T. Sundius, MOLVIB: a program for harmonic force field calculation, QCPE Program No. 604 (1991)

Chapter 2
Structural and Vibrational Study of Chromyl Fluorosulfate

Abstract In this chapter we present a structural and vibrational study related to chromyl fluorosulfate. The compound was prepared and characterized by infrared spectroscopy. The density functional theory (DFT) has been used to study its structure and vibrational properties. The molecular structure of the compound has been theoretically determined in gas phase employing the B3LYP, B3P86, and B3PW91 levels of theory, and the harmonic vibrational frequencies were evaluated at the same levels. The calculated harmonic vibrational frequencies for chromyl fluorosulfate are consistent with the experimental IR spectrum. These calculations gave us a precise knowledge of the normal modes of vibration taking into account the type of coordination adopted by fluorosulfate groups of this compound as monodentate and bidentate ligands. Also, the assignment of all the observed bands in the IR spectrum for chromyl fluorosulfate was performed. The nature of the Cr–O and Cr ← O bonds and the topological properties of the compound were investigated and analyzed by means of natural bond order (NBO) and *Bader's* Atoms in Molecules theory (AIM), respectively.

Keywords Chromyl fluorosulfate · Vibrational spectra · Molecular structure · Force field · DFT calculations

2.1 Introduction

Since long time, the compounds that contain the V and Cr atoms [1–13] have been studied because many of them present interesting properties, such as the vanadium oxo [2–4, 6, 8, 9, 12] and chromyl compounds [1, 4, 5, 11, 13]. Thus, the chromyl

Adapted from Journal of Molecular Structure,981/1–3, A. Ben Altabef, S.A. Brandán, A New Vibrational Study of Chromyl Fluorosulfate, CrO2 (SO3F)2 by DFT calculations, 146–152, copyright 2010, with permission from Elsevier

nitrate and perchlorate compounds were theoretical and recently studied by means of the normal mode calculations accomplished by use of a generalized valence force field (GVFF) in order to analyze the coordination modes of the nitrate and perchlorate groups and carry out their complete assignments [11, 13]. These groups can act as monodentate or bidentate ligand [14–19].

The experimental molecular structure of chromyl nitrate has C_2 symmetry [20], while two structures denominate, $C_2(1)$ and $C_2(2)$, were theoretically found for chromyl perchlorate, both with C_2 symmetries [13]. For the $C_2(1)$ structure the two coordination modes are possible, whereas for the $C_2(2)$ structure the perchlorate groups can act only as monodentate ligand. For both molecules, we demonstrate that a molecular force field considering the nitrate and perchlorate groups as monodentate and bidentate ligand calculated using the DFT/B3LYP calculations are well represented [11, 13]. Also, the chromyl fluorosulfate compound, $CrO_2(SO_3F)_2$, presents vibrational properties imperfectly described and only the main characteristics of the infrared spectrum were published in solid phase [21–23].

Initially, this compound was reported by *Lustig* and *Cady* [21] as a dark brown, slightly volatile solid that can be obtained by several methods and it is decomposed at room temperature into a greenish compound. Then, *Rochat* and *Gard* [22] report this compound as a green product relatively stable and nonvolatile at room temperature. When the moss green is heated to 75 °C, it results a stable brown solid. However, the infrared spectra of both solid compounds are the same but; the lines were much sharper and more clearly defined in the brown compound. In this case, we have prepared this compound according to *Brown* and *Gard* [23] and the obtained product was a green solid. The aim of this chapter is to perform an experimental and theoretical study on this compound with the methods of quantum chemistry in order to know its vibrational properties and carry out its complete assignment. In this case, the normal mode calculations were accomplished using a GVFF and considering the fluorosulfate group as monodentate and bidentate ligand. For that purpose, the optimized geometry and frequencies for the normal modes of vibration were calculated. Then, the performed calculations were used to predict the Raman spectrum for which no experimental data exist. For chromyl fluorosulfate, two structures, both with C_2 symmetries, were obtained in similar form to chromyl perchlorate [13]. Bell et al. [24] have found that for chromium, oxo anions and oxyhalide compounds, the B3LYP/Lanl2DZ combination gives the best fit for the geometries and the observed vibrational spectra. In this case, for the two structures of chromyl fluorosulfate, the B3LYP/6-31G* method was used. The force field for both structures of the compound was obtained using the transferable scaling factors of *Rauth* and *Pulay* [25–27] and those scaling factors obtained from chromyl nitrate [11]. Density functional theory (DFT) normal mode assignments, in terms of the potential energy distribution, are in general accord with those obtained from the normal coordinate analysis. In addition, the natural bond order (NBO) [28–31] and atoms in molecules (AIM) [32, 33] calculations were performed in order to study the nature of the two types of Cr–O and Cr ← O bonds and the topological properties of electronic charge density, respectively.

2.2 Structural Study

For chromyl fluorosulfate, using the different methods and basis sets, two different structures were found, as in chromyl perchlorate, both with C_2 symmetries named $C_2(1)$ and $C_2(2)$. In the first structure, the fluorosulfate groups can act as a monodentate or bidentate ligand (Fig. 2.1) while in the second one, the SO_3F^- groups can act only as monodentate ligands (Fig. 2.2).

Table 2.1 shows the comparison of the total energies and dipole moment values for both structures of chromyl fluorosulfate by using the Lanl2dz, STO-3G*, 3-21G*, 6-31G, 6-31G*, 6-311+G, 6-311 ++G, and 6-311 ++G** basis sets with the B3LYP method, while for the B3P86 and B3PW91 methods only the STO-3G* and 6-31G* basis sets were used [34–39].

Note that for the $C_2(1)$ structure, by using the B3LYP method, many basis sets have imaginary frequency values, as also with the B3PW91/6-311G** method, while for the $C_2(2)$ structure all the frequency values were positive. Moreover, the lower energy values for both structures were obtained by using the B3P86/6-311G** combination, while a low energy difference (ΔE) among both structures (0.52 kJ/mol) was obtained by using B3P86/6-31G*, as shown in Table 2.1. A comparison of experimental data for chromyl nitrate with the calculated

Fig. 2.1 The $C_2(1)$ molecular structure of chromyl fluorosulfate considering the chlorosulfate group as: **a** monodentate ligand and **b** bidentate ligand. Reprinted from Journal of Molecular Structure, 981/1–3, A. Ben Altabef, S.A. Brandán, a new vibrational study of chromyl fluorosulfate, CrO2 (SO3F)2 by DFT calculations 146–152, copyright 2010, with permission from Elsevier

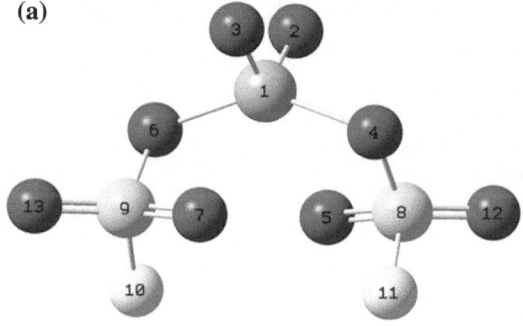

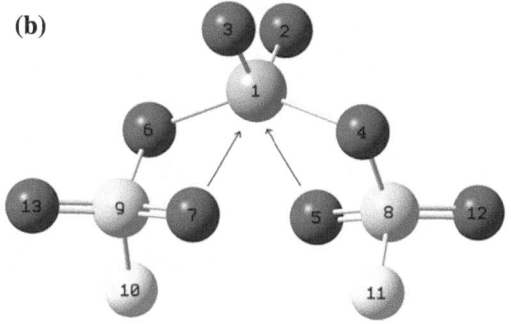

Fig. 2.2 The $C_2(2)$
molecular structure of
chromyl fluorosulfate
considering the chlorosulfate
group as monodentate ligand.
Reprinted from Journal
of Molecular Structure,
981/1–3, A. Ben Altabef,
S.A. Brandán, a new
vibrational study of chromyl
fluorosulfate, CrO2 (SO3F)2
by DFT calculations 146–
152, copyright 2010, with
permission from Elsevier

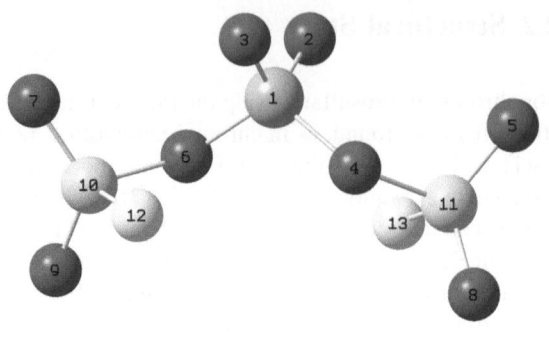

Table 2.1 Total energy (ET) and dipole moment () for two structures of chromyl fluorosulfate using different theory levels

B3LYP method						
$C_2(1)$ Symmetry			$C_2(2)$ Symmetry			ΔE
Basis set	ET (Hartree)	μ (D)	Basis set	ET (Hartree)	μ (D)	kJ/mol
LanL2DZ[a]	−907.5075	1.46	LanL2DZ	−907.5292	1.18	56.92
6-31G[a]	−2641.5519	0.84	6-31G	−2641.5711	0.93	50.36
6-31G*	−2642.1638	0.63	6-31G*	−2642.1671	0.30	8.65
6-31G**	−2642.1638	0.63	6-31G**	−2642.1672	0.30	8.92
6-311G***[a]	−2642.4855	1.48	6-311G**	−2642.4980	1.26	32.78
6-311 + G[a]	−2641.9696	1.62	6-311 + G	−2641.9905	0.07	54.82
6-311 ++G[a]	−2641.9696	1.63	6-311 ++G	−2641.9905	0.07	54.82
6-311 ++G***[a]	−2642.5176	1.40	6-311 ++G**	−2642.5313	1.81	35.93
B3P86 method						
6-31G*	−2644.7194	0.53	6-31G*	−2644.7192	0.29	0.52
6-31G**	−2644.7194	0.53	6-31G**	−2644.7192	0.29	0.52
6-311G**	−2645.0283	1.27	6-311G**	−2645.0354	1.00	18.62
B3PW91 method						
6-31G*	−2641.7296	0.55	6-31G*	−2641.7159	0.57	35.93
6-31G**	−2641.7143	0.58	6-31G**	−2641.7158	0.29	3.93
6-311G***[a]	−2642.0215	1.33	6-311G**	−2642.0303	0.95	23.08

[a] Imaginary frequencies

geometrical parameters for the $C_2(1)$ and $C_2(2)$ structures of chromyl fluorosulfate by using a 6-31G* basis set at different theory levels can be seen in Table 2.2. According to these results, the methods that best reproduce the experimental geometrical parameters for chromyl fluorosulfate with $C_2(1)$ structure is B3PW91/ 6-31G* where the mean difference for bond lengths is 0.050 Å, while with the B3LYP/6-31G* method it is 5.57° for angles. The functional B3P86 shows a somewhat less satisfactory agreement (5.74°). On the other hand, in the $C_2(2)$

Table 2.2 Comparison of experimental and calculated geometrical parameters at different theory levels for both structures of chromyl fluorosulfate.

$CrO_2(SO_3F)_2$							$CrO_2(NO_3)_2$
Parameter	$C_2(1)$ Symmetry			$C_2(2)$ Symmetry			C_2 Symmetry
	B3LYP 6-31G*	B3PW91	B3P86	B3LYP 6-31G*	B3PW91	B3P86	Ref. [20]
Bond length (Å)							
R(1,2)	1.544	1.541	1.536	1.544	1.536	1.535	1.586 (2)
R(1,3)	1.544	1.541	1.536	1.544	1.536	1.535	1.586 (2)
R(1,4)	1.910	1.913	1.902	1.785	1.778	1.775	1.957 (5)
R(1,5)	2.369	2.315	2.319	3.274	3.264	3.216	2.254 (20)
R(1,6)	1.910	1.913	1.902	1.785	1.778	1.775	1.957 (2)
R(1,7)	2.369	2.315	2.319	3.274	3.264	3.216	2.254 (20)
R(4,8)	1.562	1.551	1.551	1.623	1.614	1.611	
R(5,8)	1.478	1.476	1.474	1.442	1.438	1.437	
R(6,9)	1.562	1.551	1.551	1.623	1.614	1.611	
R(7,9)	1.478	1.476	1.474	1.442	1.438	1.437	
R(8,11)	1.587	1.581	1.577	1.439	1.584	1.581	
R(8,12)	1.436	1.431	1.430	1.439	1.435	1.433	
R(9,10)	1.587	1.581	1.577	1.590	1.584	1.581	
R(9,13)	1.436	1.431	1.430	1.590	1.435	1.433	
RMSD	0.075	0.050	0.057	0.597	0.592	0.566	
Bond angle (°)							
A(2,1,3)	106.7	106.5	106.6	109.1	108.9	108.9	112.2 (71)
A(2,1,4)	103.6	103.4	103.5	110.3	110.3	110.3	97.2 (18)
A(2,1,5)	90.5	90.4	90.5	75.3	75.2	74.5	
A(2,1,6)	97.8	97.2	97.5	107.6	107.6	107.5	104.5 (9)
A(2,1,7)	159.4	159.6	159.6	154.1	153.8	154.7	
A(3,1,4)	97.8	97.2	97.5	107.6	107.6	107.5	104.5 (9)
A(3,1,5)	159.4	159.6	159.6	154.1	153.8	154.7	
A(3,1,6)	103.6	103.4	103.5	110.3	110.3	110.3	97.2 (18)
A(3,1,7)	90.5	90.4	90.5	75.3	75.2	74.5	
A(4,1,5)	66.4	67.2	67.1	49.3	49.2	50.3	
A(4,1,6)	143.7	145.2	144.6	111.8	112.1	112.2	140.5 (9)
A(4,1,7)	84.7	85.1	84.7	91.6	92.1	91.6	
A(5,1,6)	84.7	85.1	84.7	91.6	92.1	91.6	
A(5,1,7)	75.5	75.8	75.4	112.3	112.9	113.6	82.8 (60)
A(6,1,7)	66.4	67.2	67.1	49.3	49.2	50.3	
A(1,4,8)	103.5	101.9	102.5	126.8	126.9	126.1	
A(1,5,8)	87.4	87.9	87.9	68.2	68.1	69.2	
A(1,6,9)	103.5	101.9	102.5	126.8	126.9	126.1	
A(1,7,9)	87.4	87.9	87.9	68.2	68.1	69.2	
A(4,8,5)	102.6	102.7	102.5	109.5	109.6	109.5	
RMSD	5.57	5.90	5.74	13.05	12.96	12.92	

Reprinted from Journal of Molecular Structure, 981/1–3, A. Ben Altabef, S.A. Brandán, a new vibrational study of chromyl fluorosulfate, CrO2 (SO3F)2 by DFT calculations146–152, copyright 2010, with permission from Elsevier

structure, the methods that best reproduce the experimental geometrical parameters for chromyl fluorsulfate is B3P86/6-31G* where the mean difference for bond lengths is 0.566 Å and for angles is 12.92° with the B3PW916-31G* method.

An important observation in the $C_2(2)$ structure is the calculated low value of the O4-Cr1-O6 bond angle with all used methods (between 111.8 and 112.2°) in relation to the experimental value of chromyl nitrate (140.4°). Table 2.3 shows the comparison of the total energies and dipole moment values for fluorosulfate ion with C_{3v} symmetry by using a 6-31G* basis set at different theory levels. Here, the structure with lower energy and dipole moment values by using B3LYP/6-31G* calculation are obtained.

A comparison of the calculated geometrical parameters by using 6-31G* basis set at different theory levels for the fluorosulfate ion to the corresponding experimental values of the SOF_2, SO_2F_2, SO_3 [40], and $LiSO_3F$ [41] compounds in Table 2.4 is shown. These results reveal that the method that best reproduce the

Table 2.3 Total energy (*ET*) and dipole moment () for SO_3F^- ion using 6-31G* basis set at different theory level

Method	ET (Hartree)	μ (D)
MP2	−722.3955	0.91
B3PW91	−723.5550	0.85
B3LYP	−723.7217	0.82
B3P86	−724.6294	0.86

Table 2.4 Comparison of experimental and calculated geometrical parameters for SO_3F^- ion using 6-31G* basis set at different theory levels

Parameters	[a]Theoretical					Experimental			
	Ab initio HF	DFT methods			PostHF	[b]SOF_2	[b]SO_2F_2	[b]SO_3	[c]$LiSO_3F$
		B3LYP	B3PW91	B3P86	MP2				
Bond length (Å)									
R(1,2)	1.602	1.670	1.662	1.658	1.667	1.585	1.570		1.555 (7)
R(1,3)	1.436	1.473	1.469	1.467	1.471	1.412	1.370	1.430	1.455 (6)
R(1,4)	1.436	1.473	1.469	1.467	1.471			1.430	1.424 (4)
R(1,5)	1.436	1.473	1.469	1.467	1.471			1.430	1.424 (4)
Bond angle (°)									
A(2,1,3)	102.4	102.1	102.1	102.1	101.9			120	104.5 (5)
A(2,1,4)	102.4	102.1	102.1	102.1	101.9			120	102.8 (3)
A(2,1,5)	102.4	102.1	102.1	102.1	101.9			120	
A(3,1,4)	115.5	115.7	115.7	115.7	115.8				117.4 (4)
A(3,1,5)	115.5	115.7	115.7	115.7	115.8				113.5 (2)
A(4,1,5)	115.5	115.7	115.7	115.7	115.8				

[a] This work
[b] Ref. [40]
[c] Ref. [41]

experimental distances for fluorosulfate ion is HF where the mean difference for bond lengths is 0.014 Å (related to SOF_2), 0.036 Å (related to SO_2F_2), and 0.073 Å (related to $LiSO_3F$). This greater variation in the distance in $LiSO_3F$ is justified, because the SO_3F^- group forms a slightly distorted tetrahedron with a fixed position for the F atom and the S–F in other direction. On the order hand, the theoretical bond angles in all cases are closer to the values for the $LiSO_3F$ compound. The calculated S–O and S–F bond lengths for the fluorosulfate groups corresponding to the $C_2(1)$ structure of chromyl fluorosulfate, by using the three studied methods, are those that better reproduce the experimental geometrical parameters of the SO_3F^- group (related to $LiSO_3F$) with a mean difference of 0.052 Å for distances and 5.5° for angles, while these values for the $C_2(2)$ structure are respectively, 0.090 Å and 6.9°.

The bond orders, expressed by Wiberg's indexes, for both structures by using a B3LYP/6-31G* calculation are shown in Table 2.5. In the $C_2(1)$ structure, the chromium atom forms six bonds, two Cr=O bonds (bond order 2.0077), two Cr–O (bond order 0.4844), and two Cr ← O (bond order 0.1196), while with the other methods the values slightly change. For the $C_2(2)$ structure, the Cr atom forms only four bonds (Table 2.5) because the bond order values for the two Cr ← O bonds change at 0.0158. The DFT calculations predict for both structures that the O4-Cr1-O6 angles are higher than the O_2=Cr1=O_3 bond angle in accordance with the results obtained from chromyl nitrate [20] and perchlorate [13]. This contradiction with the Valence-Shell Electron-Pair Repulsion (VSEPR) theory [42, 43] could be explained in a way similar to other compounds by means of molecular orbital (MO) studies by analyzing the delocalized and/or bonding characters of the relevant MO [7, 11, 13]. The intermolecular interactions for the $C_2(1)$ structure have been analyzed by using *Bader's* topological analysis of the charge electron density, $\rho(r)$ using the AIM program [33].

For the characterization of molecular electronic structure it is important to determine the $\rho(r)$ in the bond critical points (BCPs) and the values of the Laplacian at these points.

The analyses of the Cr–O and Cr ← O BCPs for the $C_2(1)$ structure with the B3LYP/6-31G* and B3P86/6-31G* methods are reported and compared with the corresponding bidentate structure for chromyl perchlorate by using a B3P86/6-31G* level in Table 2.6. Here, there are two important observations, in one case, the Cr1 ← O5 and Cr1 ← O7 BCP have the typical properties of the closed-shell interaction ($\rho(r) = 0.04$ a.u., $|\lambda 1|/\lambda 3 < 1$, and $\nabla^2\rho(r) = 0.20$ a.u.) [44] while the other important observation is related to the topological properties of the Cr1-O4 and Cr1−O6 BCPs, since in both cases they are the same. It is important to note that the properties of the Cr–O and Cr ← O BCPs in chromyl perchlorate are slightly higher than the corresponding values of chromyl fluorosulfate. This difference is related principally with the used method because the properties of chromyl fluorosulfate are closer to chromyl perchlorate when the calculation is performed with the B3P86 method. On the other hand, the (3, +1) critical points confirm the two Cr ← O bonds in the $C_2(1)$ structure of chromyl fluorosulfate (Table 2.6). These two ring points reveal that the coordination mode adopted for

Table 2.5 Wiberg index bond matrix for both structures of chromyl fluorosulfate

B3LYP/6-31G* method

$C_2(1)$ Symmetry

Atoms	1	2	3	4	5	6	7	8	9	10	11	12	13
1 Cr	0.0000	2.0077	2.0077	0.4844	0.1196	0.4844	0.1196	0.0135	0.0135	0.0078	0.0078	0.0294	0.0294
2 O	2.0077	0.0000	0.2442	0.0814	0.0076	0.0619	0.0406	0.0048	0.0071	0.0011	0.0012	0.0050	0.0066
3 O	2.0077	0.2442	0.0000	0.0619	0.0406	0.0814	0.0076	0.0071	0.0048	0.0012	0.0011	0.0066	0.0050
4 O	0.4844	0.0814	0.0619	0.0000	0.0838	0.0235	0.0046	0.8908	0.0022	0.0012	0.0516	0.0938	0.0014
5 O	0.1196	0.0076	0.0406	0.0838	0.0000	0.0046	0.0034	1.1840	0.0021	0.0009	0.0753	0.1314	0.0008
6 O	0.4844	0.0619	0.0814	0.0235	0.0046	0.0000	0.0838	0.0022	0.8908	0.0516	0.0012	0.0014	0.0938
7 O	0.1196	0.0406	0.0076	0.0046	0.0034	0.0838	0.0000	0.0021	1.1840	0.0753	0.0009	0.0008	0.1314
8 S	0.0135	0.0048	0.0071	0.8908	1.1840	0.0022	0.0021	0.0000	0.0003	0.0005	0.6575	1.4362	0.0005
9 S	0.0135	0.0071	0.0048	0.0022	0.0021	0.8908	1.1840	0.0003	0.0000	0.6575	0.0005	0.0005	1.4362
10 F	0.0078	0.0011	0.0012	0.0012	0.0009	0.0516	0.0753	0.0005	0.6575	0.0000	0.0001	0.0002	0.0949
11 F	0.0078	0.0012	0.0011	0.0516	0.0753	0.0012	0.0009	0.6575	0.0005	0.0001	0.0000	0.0949	0.0002
12 O	0.0294	0.0050	0.0066	0.0938	0.1314	0.0014	0.0008	1.4362	0.0005	0.0002	0.0949	0.0000	0.0001
13 O	0.0294	0.0066	0.0050	0.0014	0.0008	0.0938	0.1314	0.0005	1.4362	0.0949	0.0002	0.0001	0.0000

$C_2(2)$ Symmetry

Atoms	1	2	3	4	5	6	7	8	9	10	11	12	13
2 O	1.8991	0.0000	0.2537	0.0851	0.0101	0.1051	0.0018	0.0103	0.0053	0.0048	0.0096	0.0032	0.0019
3 O	1.8991	0.2537	0.0000	0.1051	0.0018	0.0851	0.0101	0.0053	0.0103	0.0096	0.0048	0.0019	0.0032
4 O	0.6909	0.0851	0.1051	0.0000	0.1153	0.0516	0.0012	0.0994	0.0020	0.0012	0.6796	0.0007	0.0618
5 O	0.0158	0.0101	0.0018	0.1153	0.0000	0.0012	0.0000	0.2108	0.0002	0.0001	0.0012	0.0001	0.1211
6 O	0.6909	0.1051	0.0851	0.0516	0.0012	0.0000	0.1153	0.0020	0.0994	0.6796	0.0001	0.0618	0.0007
7 O	0.0158	0.0018	0.0101	0.0012	0.0000	0.1153	0.0000	0.0002	0.2108	0.0001	0.0001	0.1211	0.0001
8 S	0.0077	0.0103	0.0053	0.0994	0.2108	0.0020	0.0002	0.0000	0.0001	0.0000	0.5915	1.3882	0.0001
9 S	0.0077	0.0053	0.0103	0.0020	0.0002	0.0994	0.2108	0.0001	0.0000	0.5915	0.0000	0.0001	1.3969
10 F	0.0073	0.0048	0.0096	0.0012	0.0001	0.6796	0.0001	0.0000	0.5915	0.0000	0.0000	0.0001	0.1210
11 F	0.0073	0.0096	0.0048	0.6796	0.0012	0.0001	0.0001	0.5915	0.0000	0.0000	0.0000	0.1210	0.0001
12 O	0.0282	0.0032	0.0019	0.0007	0.0001	0.0618	0.1211	1.3882	0.0001	0.0001	0.1210	0.0000	0.0001
13 O	0.0282	0.0019	0.0032	0.0618	0.1211	0.0007	0.0001	0.0001	1.3969	0.1210	0.0001	0.0001	0.0000

Table 2.6 Analysis of Cr ← O bond critical points in the $C_2(2)$ structure of chromyl fluorosulfate and chromyl perchlorate

[a]Chromyl fluorsulfate

B3LYP/6-31G* method

$C_2(1)$ Structure coordination bidentate

Parameter[c]	Cr1-O4	Cr1 ← O5	Cr1-O6	Cr1 ← O7	(3, +1)	(3, +1)
$\rho(r)$	0.1114	0.0359	0.1114	0.0359	0.0325	0.03256
$\nabla\rho(r)$	0.5236	0.1340	−0.2090	0.1343	0.1311	−0.0328
$\lambda 1$	−0.2095	−0.0375	−0.1900	−0.0374	−0.0309	−0.0309
$\lambda 2$	−0.1910	−0.0315	−0.1900	−0.0314	0.0401	0.0402
$\lambda 3$	0.9241	0.2030	0.9245	0.2031	0.1218	0.1219
$|\lambda 1|/\lambda 3$	0.2267	0.1847	0.2055	0.1841	0.2536	0.2535
B3P86/6-31G* method						
$\rho(r)$	0.1142	0.0403	0.1142	0.0403	0.0354	0.0354
$\nabla^2\rho(r)$	0.5380	0.1595	0.5359	0.1591	0.1475	0.1474
$\lambda 1$	−0.2150	−0.0438	−0.2156	−0.0439	−0.0343	−0.0343
$\lambda 2$	0.1959	−0.0389	−0.1969	−0.0391	0.0496	0.0495
$\lambda 3$	0.9489	0.2423	0.9484	0.2421	0.1322	0.1321
$|\lambda 1|/\lambda 3$	0.2266	0.1808	0.2273	0.1813	0.2594	0.2596
[b]Chromyl perchlorate						
B3P86 method						
$\rho(r)$	0.1191	0.0404	0.1191	0.0404	0.0339	0.03390
$\nabla^2\rho(r)$	0.5641	0.1656	0.5647	0.1663	0.1415	0.1415
$\lambda 1$	−0.2295	−0.0442	−0.2292	−0.0442	−0.0316	−0.0316
$\lambda 2$	−0.2066	−0.0422	−0.2063	−0.0418	0.0463	0.0463
$\lambda 3$	1.0003	0.2521	1.0002	0.2522	0.1268	0.1268
$|\lambda 1|/\lambda 3$	0.2294	0.1753	0.2291	0.1752	0.2492	0.2492

[a] This work; [b] Ref. [13]; [c] The quantities are in atomics units

the fluorosulfate groups in that structure is bidentate. On the other hand, for the $C_2(2)$ structure, using the B3LYP/6-31G* and B3P86/6-31G* methods the Cr ← O BCPs were not possible to be seen, and for this the coordination of the fluorosulfate groups in this structure is only possible as monodentate ligands. The characteristic of these Cr–O BCPs with all the methods used can be seen in Table 2.7. Again, the Cr1-O4 and Cr1-O6 BCPs in chromyl perchlorate are slightly higher with the two methods than the corresponding values of chromyl fluorosulfate.

2.3 Vibrational Study

The two structures of chromyl fluorosulfate have C_2 symmetry and 33 active vibrational normal modes in the infrared and Raman spectra (17 A + 16 B). In this chapter, the study was performed taking into account the monodentate and bidentate coordination modes by fluorosulfate groups, because it is impossible to

Table 2.7 Analysis of Cr–O bond critical points in the $C_2(2)$ structures of chromyl fluorosulfate and chromyl perchlorate at different theory levels

$C_2(2)$ Symmetry coordination monodentate

Parameter[c]	[a]Chromyl fluorosulfate		[b]Chromyl perchlorate	
	Cr1-O4/Cr1-O6			
	B3LYP 6-31G*	B3P86 6-31G*	B3LYP 6-31G*	B3P86 6-31G*
$\rho(r)$	0.1478	0.1522	0.1532	0.1573
$\nabla^2 \rho(r)$	0.7606	0.7915	0.7634	0.7787
$\lambda 1$	−0.2887	−0.2977	−0.3061	−0.3158
$\lambda 2$	−0.2673	−0.2762	−0.2785	−0.2874
$\lambda 3$	1.3167	1.3665	1.3480	1.3819
$|\lambda 1|/\lambda 3$	0.2192	0.2178	0.2271	0.2285

[a] This work
[b] Ref. [13]
[c] The quantities are in atomics units

make a difference between both coordination modes [18] on the grounds of infrared and Raman spectra alone. The observed frequencies and the assignment for chromyl fluorosulfate are given in Table 2.8. Vibrational assignments were made on the basis of the potential energy distributions (PED) in terms of symmetry coordinates and by comparison with molecules that contain similar groups [1, 11, 13, 22, 23, 45–50]. Here, the results obtained are related at B3LYP level with 6-31G* basis set because after scaling, by using this method, a satisfactory agreement is obtained between the calculated and the experimental vibrational frequencies of chromyl fluorosulfate. In general, the theoretical infrared spectrum of the chromyl fluorosulfate for the $C_2(1)$ structure demonstrates slight agreement with the experimental spectrum, As in Fig. 2.3, it is observed that the theoretical infrared spectrum for the $C_2(2)$ structure is slightly different from the experimental ones. It is possible to observe that in all calculations some vibrational modes of different symmetries are mixed among them because the frequencies are approximately the same. Below we discuss the assignment of the most important groups.

2.4 Coordination Monodentate of the Fluorosulfate Groups

2.4.1 $C_2(1)$ and $C_2(2)$ Symmetries

The frequencies, IR and Raman intensities and PED, obtained by B3LYP/6-31G* calculations considering this mode of coordination for both structures appear in Tables 2.9 and 2.10. Here, the covalent bonding of the fluorosulfate group is easily recognized from its infrared spectrum because the symmetry group changes from the point group C_{3v} of the free ion to C_2 of the compound.

Table 2.8 Experimental and calculated frequencies (cm^{-1}), potential energy distribution and assignment for chromyl fluorosulfate.

| | Experimental | | | | | ^{a}SQM/B3lyp/6-31G* | | | ^{a}Assignment | | |
| | | | | | | C₂(1) | | C₂(2) | C₂(1) | | C₂(2) |
Modes	^{a}IR	^{b}IR	^{c}Raman	^{d}IR	^{e}IR	M	B	M	Monodentate	Bidentate	Monodentate
A symmetry											
1	1329 s	1438 w	1429 s	1420 w	1308 s	1416	1394	1431	ν_a S=O₂ ip	ν_s S=O	ν_a S=O₂ op
2	1178 sh	1215 vs		1195 s, br		1138	1179	1194	ν_s S=O₂ op	ν_a S-O₂ op	ν_s S=O₂ ip
3	1055 sh	1161 s			1098 m	1001	1010	1117	ν_s S-O₂	ν_s S-O₂ ip	ν_a Cr=O
4	1026 s	1020 vs				961	961	920	ν_s Cr=O	ν_s Cr=O	ν_a S-O₂
5	794 m	926 s	850 (109)	810 s		846	850	820	ν_a S-F	ν_a S-F	ν_a S-F
6	666 sh			618 m	633 m	657	679	665	ν_s Cr-O	δSO₂ op	ν_s Cr-O
7	580 sh		560 (10) p?	575 m	562 m	576	562	535	δSO₂ op	ρSO₂ op	ρSO₂ op
8	534 w		555	550 s		516	509	531	τSO₂ ip	Wag SO₂ op	Wag SO₂ op
9	476 sh					470	475	465	Wag SO₂ op	δS-O-F ip	δCrO₂
10			405 (8) dp		422 m	433	445	417	δCrO₂	δCrO₂	δSO₂ ip
11						359	341	365	δS-O-F ip	τwSO₂ ip	τSO₂ op
12					262 s	273	274	251	τSO₂ op	νCr-O	δS-O-F ip
13						212	220	225	ρ CrO₂	ρCrO₂	τw CrO₂ op
14					165 m	147	148	159	δs Cr-O-S	δCrO₂ ip^f	τw CrO₂ ip
15						133	136	101	τCr-O-S-O ip	νCr-O	δa Cr-O-S
16					96 w	89	93	36	τw CrO₂ ip	τwCrO₂ ip	τO-Cr-O-S op
17						51	56	16	τO-Cr-O-S ip	τCr-O-S-O ip	τCr-O-S-O op
B symmetry											
18	1233 sh	1374 vs		1350 s	1207 s	1403	1379	1427	ν_a S=O₂ op	ν_a S=O	ν_a S=O₂ ip
19	1209 vs	1245 s	1230 (6) p			1153	1191	1198	ν_s S=O₂ ip	ν_a S-O₂ ip	ν_s S=O₂ op
20	1026 s	1061 s		1050 s, br		978	979	1102	ν_a S-O₂	ν_s S-O₂ op	ν_s Cr=O
21	989 sh	992 s	960 (6) p	955 s,br		944	946	850	ν_a Cr=O	ν_a Cr=O	ν_s S-O₂
22	871 w	948 m		910 s, br	842 m	847	853	831	ν_s S-F	ν_s S-F	ν_s S-F
23	734 m					665	684	690	ν_a Cr-O	δSO₂ ip	ν_a Cr-O

(continued)

Table 2.8 (continued)

Experimental						aSQM/B3lyp/6-31G*			aAssignment		
						C2(1)		C2(2)	C2(1)		C2(2)
Modes	aIR	bIR	cRaman	dIR	cIR	M	B	M	Monodentate	Bidentate	Monodentate
24	611 sh				607 m	584	578	567	δSO₂ ip	ρSO₂ ip	τSO₂ ip
25	514 w					512	506	506	ρSO₂ op	Wag SO₂ ip	ρSO₂ ip
26	450d			450 m, br	431 m	456	462	448	ρSO₂ ip	δS–O–F op	δSO₂ op
27			391			361	355	370	δS–O–F op	ν Cr–O	Wag CrO₂
28	330d			330 m	309 m	336	334	347	Wag SO₂ ip	τSO₂ op	Wag SO₂ ip
29						237	240	233	Wag CrO₂	Wag CrO₂	ρ CrO₂
30					177 sh	195	204	225	τO–Cr–O–S op	τwCrO₂ op	δS–O–F op
31					121 w	123	128	94	τw CrO₂ op	δCrO₂ opf	δs Cr–O–S
32						67	74	22	τCr–O–S–O op	τCr–O–S–O op	τO–Cr–O–S ip
33						10	10	11	δa Cr–O–S	ν Cr–O	τCr–O–S–O ip

Abbreviations: ν stretching, δ, deformation, ρ rocking, wag, (γ) wagging, τw torsion, a antisymmetric, s symmetric, op out-of-phase, ip in-phase, M monodentate, B bidentate, ν very, s strong, m medium, w weak, sh shoulder, br broad, p polarized, and dp depolarized

a This work
b Ref. [22] CrO₂(SO₃F)₂
c Ref. [47] HSO₃F
d Ref. [23] CrO₂(SO₃F)₂
e Ref. [45] Cu(SO₃F)₂
f O–Cr–O deformation

Reprinted from Journal of Molecular Structure, 981/1–3, A. Ben Altabef, S.A. Brandán, a new vibrational study of chromyl fluorosulfate, CrO2 (SO3F)2 by DFT calculations 146–152, copyright 2010, with permission from Elsevier

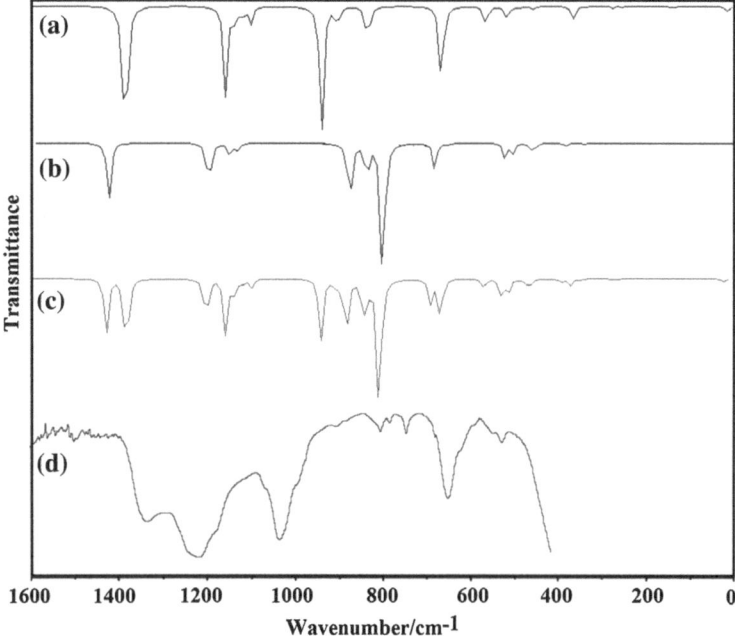

Fig. 2.3 Infrared spectrum of chromyl fluorosulfate: **a** for the theoretical $C_2(1)$ structure; **b** for the $C_2(2)$ theoretical structure; **c** calculated average infrared spectra for both structures from B3LYP/6–31G* wavenumbers and intensities using Lorentzian band shapes (for a population relation $C_2(1)$: $C_2(2)$ of 1: 1 for each structure, and **d** infrared experimental spectrum. Reprinted from Journal of Molecular Structure, 981/1–3, A. Ben Altabef, S.A. Brandán, a new vibrational study of chromyl fluorosulfate, $CrO_2 (SO_3F)_2$ by DFT calculations146–152, copyright 2010, with permission from Elsevier

2.4.2 Fluorosulfate Groups

The IR bands observed in $CrO_2(SO_3F)_2$ in the 1374–1061 cm^{-1} region were assigned by Rochat and Gard [22] to the two $S=O_2$ stretching modes, as in the $Cu(SO_3F)_2$ compound [45] (1308 and 1207 cm^{-1}). In this chapter, the $S=O_2$ in-phase and out-of-phase antisymmetric stretching modes for the $C_2(1)$ structure are assigned to the IR band and shoulder at 1329 and 1233 cm^{-1} respectively, while these last two bands, in the $C_2(2)$ structure, are associated with the $S=O_2$ out-of-phase and in-phase antisymmetric modes, respectively. The corresponding $S=O_2$ out-of-phase and in-phase symmetric stretching modes for the $C_2(1)$ structure are assigned to the very strong band and the shoulder at 1209 and 1178 cm^{-1}, respectively, while in the $C_2(2)$ structure, both bands are associated in an inverse relation with those modes. The SO_2 symmetric and antisymmetric stretching modes are predicted at 1001 and 978 cm^{-1} respectively, for this, these modes in the $C_2(1)$ structure are associated to the shoulder at 1055 cm^{-1} and to the strong band at 1026 cm^{-1}. Experimentally, in the fluorosulfate compounds, the S–F stretching modes are observed between 890 and 700 cm^{-1} [22, 23, 45–50].

Table 2.9 Experimental and calculated frequencies (cm^{-1}), potential energy distribution and assignment for C$_2$(1) Structure of chromyl fluorosulfate

C$_2$(1) Structure monodentate coordination

Modes	Observed[a]	Calculated[b]	SQM[c]	PED ($\geq$10 %)
A symmetry				
1	1329 s	1387	1416	ν_a SO$_2$ ip (75)
2	1178 sh	1127	1138	ν_s SO$_2$ ip (68) + ν_a SO$_2$ op (15)
3	1055 sh	1109	1001	ν_s SO$_2$ (65) + ρ SO$_2$ op (12)
4	1026 s	931	961	ν_s Cr=O (95)
5	794 m	826	846	ν S–F op (77)
6	666 sh	657	657	wag SO$_2$ op (34) + ν_s Cr–O (32) + δ Cr–O–S ip (13) + τSO$_2$ ip (11)
7	580 sh	557	576	τSO$_2$ ip (47) + δ SO$_2$ op (23) + ρ SO$_2$ ip (18)
8	534 w	556	516	τSO$_2$ ip (44) + δ SO$_2$ op (14) + τ O–Cr–O–S op (11) + ν S–F op (10)
9	476 sh	478	470	wag SO$_2$ op (30) + τSO$_2$ ip (25) + δCrO$_2$ (23)
10		436	433	δ CrO$_2$ (40) + wag SO$_2$ op (17) + τ O–Cr–O–S ip (12)
11		347	359	δ O–S–F ip (37) + δ SO$_2$ ip (22) + τSO$_2$ op (20) + τ O–Cr–O–S ip (10)
12		270	273	ν_s Cr–O (31) + τSO$_2$ op (27) + δ O–S–F ip (11)
13		235	212	ρ CrO$_2$ (65) + τ O–Cr–O–S ip (12)
14		155	147	δ Cr–O–S ip (36) + ν_s Cr–O (14) + wag SO$_2$ op (12) + τSO$_2$ ip (12) + τw CrO$_2$ ip (12)
15		143	133	τ Cr–O–S–O ip (52) + δ Cr–O–S ip (15) + ρ CrO$_2$ (14) + τw CrO$_2$ ip (11)
16		98	89	τ O–Cr–O–S ip (56) + τw CrO$_2$ ip (31)
17		57	51	τ Cr–O–S–O ip (49) +τ O–Cr–O–S ip (47)
B symmetry				
18	1233 sh	1374	1403	ν_a SO$_2$ op (76)
19	1209 vs	1149	1153	ν_s SO$_2$ ip (66) + ν_a SO$_2$ op (16)
20	1026 s	1091	978	ν_a SO$_2$ (62) + ν_a Cr=O (20)
21	989 sh	895	944	ν_a Cr=O (70) + ν_a SO$_2$ (22)
22	871 w	826	847	ν_a S–F (77)

(continued)

Table 2.9 (continued)

C$_2$(1) Structure monodentate coordination

Modes	Observed[a]	Calculated[b]	SQM[c]	PED (≥10 %)
23	734 m	657	665	ν_a Cr–O (41) + ρ SO$_2$ ip (22) + τSO$_2$ ip (18)
24	611 sh	557	584	δSO$_2$ ip (40) + ρ SO$_2$ op (27) + wag SO$_2$ op (10)
25	514 w	506	512	ρ SO$_2$ op (44) + τSO$_2$ ip (35)
26	450[d]	449	456	ρ SO$_2$ ip (49) + τSO$_2$ ip (15) + δSO$_2$ op (13)
27		361	361	δ O–S–F op (44) + δSO$_2$ op (19) + wag CrO$_2$ (12) + ν_a Cr–O (11)
28	330[d]	336	336	wag SO$_2$ ip (58) + ν_a Cr–O (11)
29		257	237	wag CrO$_2$ (68) + δ O–S–F op (14)
30		214	195	τw CrO$_2$ op (53) + τ O–Cr–O–S op (13)
31		132	123	τw CrO$_2$ op (29) + δ Cr–O–S op (21) + τ O–Cr–O–S op (14) + τ Cr–O–S–O op (10)
32		75	67	τ Cr–O–S–O op (74) + τw CrO$_2$ op (12) + τ O–Cr–O–S op (10)
33		10	10	δ Cr–O–S op (25) + τ O–Cr–O–S op (16) + ρ SO$_2$ ip (15)

[a] This work
[b] DFT B3LYP/6-31G*
[c] From scaled quantum mechanics force field
[d] From Ref. [23]

Table 2.10 Experimental and calculated frequencies (cm^{-1}), potential energy distribution, and assignment for C$_2$(2) structure of chromyl fluorosulfate

C$_2$(2) Structure monodentate coordination

Modes	Observed[a]	Calculated[b]	SQM[c]	PED ($\geq$10 %)
A symmetry				
1	1329 s	1421	1431	v_a SO$_2$ ip (57) + ρ SO$_2$ ip (11) + τSO$_2$ ip (11)
2	1178 sh	1194	1194	v_s SO$_2$ ip (89)
3	1055 sh	1148	1117	v_a Cr=O (97)
4	1026 s	873	920	v_a SO$_2$ (55) + v_s Cr–O (23) + ρ SO$_2$ op (11)
5	794 m	797	820	v S–F op (66) + v_s SO$_2$ (13)
6	666 sh	652	665	τSO$_2$ ip (24) + v_s Cr–O (21) + ρ SO$_2$ op (16)
7	580 sh	520	535	ρ SO$_2$ ip (49) + wag SO$_2$ op (29)
8	534 w	502	531	wag SO$_2$ op (32) + ρ SO$_2$ op (29) + ρ SO$_2$ ip (22)
9	476 sh	468	465	δCrO$_2$ (59) + δSO$_2$ ip (23)
10		420	417	δSO$_2$ ip (28) + δ CrO$_2$ (25) + τSO$_2$ op (14)
11		362	365	τSO$_2$ op (41) + δSO$_2$ ip (14) + wagSO$_2$ ip (12) + twCrO$_2$ ip (10) + δO–S–F ip (10)
12		259	251	δ O–S–F ip (55) +v_s Cr–O (13) + δSO$_2$ ip (12)
13		232	225	τw CrO$_2$ op (64)
14		157	159	τw CrO$_2$ ip (60) + τSO$_2$ op (10)
15		100	101	δ_a O–S–Cr (61) + τ O–Cr–O–S op (18) + tw CrO$_2$ op (10)
16		37	36	τ O–Cr–O–S op (61) +δ_a O–S–Cr (15) + τ Cr–O–S–O op (13)
17		17	16	τ Cr–O–S–O op (67) +τ O–Cr–O–S op (17)
B symmetry				
18	1233 sh	1421	1427	v_a SO$_2$ ip (65) + v_a SO$_2$ op (21)
19	1209 vs	1197	1198	v_s SO$_2$ op (87)
20	1026 s	1133	1102	v_s Cr=O (95)
21	989 sh	837	850	v_s SO$_2$ (55) + v_a Cr–O (33)
22	871 w	831	831	v_a S–F (54) + τSO$_2$ ip (16) + ρ SO$_2$ op (16)
23	734 m	678	690	v_a Cr–O (35) + ρ SO$_2$ ip (22) + v_s SO$_2$ (19) +ρ SO$_2$ ip (16)
24	611 sh	521	567	τSO$_2$ ip (57) + ρ SO$_2$ ip (32)
25	514 w	501	506	ρ SO$_2$ ip (62)
26	450[d]	456	448	δSO$_2$ op (43) + ρ SO$_2$ ip (20) + wag SO$_2$ ip (12) + v_a Cr–O (12)
27		382	370	wag CrO$_2$ (31) + δ O–S–F op (29) + δSO$_2$ op (29)
28	330[d]	340	347	wag SO$_2$ ip (70)
29		232	233	ρCrO$_2$ (81)
30		224	225	δ O–S–F op (50) + wagCrO$_2$ (26) + v_a Cr–O (11)
31		92	94	δs Cr–O–S (86)
32		22	22	τ O–Cr–O–S ip (79) + δs Cr–O–S (10)
33		11	11	δ Cr–O–S ip (58) +τ O–Cr–O–S ip (33)

[a] This work
[b] DFT B3LYP/6-31G*
[c] From scaled quantum mechanics force field
[d] From Ref. [23]

Here, for both structures, those modes are assigned at 871 and 794 cm^{-1}, respectively. For the C$_2$(1) structure, the SO$_2$ in-phase and out-of-phase deformation modes are assigned to the shoulders at 611 and 580 cm^{-1} respectively, while these bands in the C$_2$(2) structure are attributed to the SO$_2$ in-phase torsion and out-of-phase rocking modes, respectively. The SO$_2$ wagging, rocking, and twisting modes are observed in the low frequencies region [49, 50]; thus, in the C$_2$(1) structure, the SO$_2$ out-of-phase wagging is clearly assigned at 476 cm^{-1}, while the corresponding in-phase mode is assigned, in both structures, at 330 cm^{-1} [23]. For the C$_2$(1) structure, both rocking modes are calculated with B symmetry and assigned at 514 (out-of-phase) and 450 cm^{-1} (in-phase), while for the other structure these bands are assigned to the in-phase rocking and out-of-phase deformation modes of the SO$_2$ groups, respectively. The IR band at 450 cm^{-1} is observed in the spectrum obtained by Brown and Gard [23]. The weak band at 534 cm^{-1} is assigned to the SO$_2$ in-phase twisting mode of the C$_2$(1) structure while the corresponding out-of-phase mode is associated with the band at 273 cm^{-1}. In the C$_2$(2) structure, the band at 534 cm^{-1} is assigned to the SO$_2$ out-of-phase wagging mode while the theoretical band at 251 cm^{-1} is associated with the in-phase S–O–F deformation.

2.4.3 Chromyl Group

In the C$_2$(1) structure, the Cr=O antisymmetric and symmetric stretchings are split by more than 17 cm^{-1} while that split is of 15 cm^{-1} in the C$_2$(2) structure, indicating in both cases a slight contribution of the Cr central atom in these vibrations. In the chromyl compounds, these modes appear in the 1,050–900 cm^{-1} region [13, 49, 50]; for this, the intense IR band at 1,026 cm^{-1} is assigned to Cr=O symmetric mode and the shoulder at 989 cm^{-1} is assigned to the corresponding antisymmetric mode of the C$_2$(1) structure. In the C$_2$(2) structure, both modes are assigned respectively at 1055 and 1026 cm^{-1}. Despite the fact that the split between both modes is greater in the C$_2$(2) structure (25 cm^{-1}), both modes are assigned at the same wavenumbers because the O4–Cr1–O6 bond angle is slightly lower in the C$_2$(2) structure. Thus, for both structures, the band at 734 cm^{-1} and the shoulder at 666 cm^{-1} are associated to these stretching modes, respectively. The CrO$_2$ bending mode is assigned, as the theoretical calculations predict, at 433 cm^{-1} in the C$_2$(1) structure, while for the other structure it is assigned to the shoulder located at 476 cm^{-1}. The calculations predict the wagging, rocking, and twisting modes of the CrO$_2$ group in the low frequencies region and coupled with other modes of the fluorosulfate groups. The assignment of those modes for both structures is different from each other, as observed in Tables 2.9 and 2.10, and moreover both structures are assigned according to the calculations, because it was not possible to observe bands in this region. Thus, for the C$_2$(1) structure, the wagging CrO$_2$ mode is calculated at a higher frequency (237 cm^{-1}) than the rocking mode (212 cm^{-1}) while the twisting mode is calculated at 123 cm^{-1}. On the other hand, in the $C_2(2)$ structure, the rocking mode is calculated (233 cm^{-1}) at lower frequency than the CrO$_2$ wagging mode (347 cm^{-1}) while the twisting mode is calculated at 159 cm^{-1}.

2.5 Coordination Bidentate of the Fluorosulfate Groups

2.5.1 $C_2(1)$ Symmetry

The observed and calculated IR frequencies and potential energy distribution obtained by B3LYP/6-31G* calculations considering the bidentate coordination appear in Table 2.11.

2.5.2 Fluorosulfate Groups

In this case, there are slight changes in the PED contribution and in the coupling of the modes. Here, the S=O symmetric stretching mode is calculated at higher wavenumbers than the corresponding antisymmetric mode; for this, the strong IR band at 1329 cm^{-1} and the shoulder at 1233 cm^{-1} is assigned, respectively, to these modes. The assignment of the bands associated to the SO$_2$ out-of-phase and in-phase antisymmetric and symmetric modes is similar to the monodentate type, as shown in Table 2.8, while the SO$_2$ in-phase and out-of-phase deformation modes are assigned, respectively, to the bands at 734 and 666 cm^{-1}. The SO$_2$ out-of-phase and in-phase wagging modes are assigned to the weak bands at 534 and 514 cm^{-1}, respectively. Both rocking modes are calculated with different symmetry and assigned to the shoulders at 580 (out-of-phase) and 611 cm^{-1} (in-phase), respectively. Finally, the band at 330 cm^{-1} is associated with SO$_2$ out-of-phase twisting while the corresponding in-phase mode is assigned according to calculations at 341 cm^{-1}.

2.5.3 Chromyl Group

Here, two Cr=O stretching modes and four Cr–O stretching modes are expected due to bidentate coordination of the fluorosulfate groups (Fig. 1.b). In chromyl compounds, the Cr=O stretching is observed in the 1050–900 cm^{-1} region [4, 12, 13, 47, 50]; for this reason, these modes are easily assigned (Table 2.8). One important observation is that the CrO$_2$ bending (O=Cr=O) mode appears at the same wavenumbers that in the monodentate case. Hence, it is possible to observe this mode at 445 cm^{-1} (56 %) while in chromyl nitrate it is observed at 475 cm^{-1}

Table 2.11 Experimental and calculated frequencies (cm^{-1}), potential energy distribution and assignment for $C_2(1)$ Structure of chromyl fluorosulfate

$C_2(1)$ Structure bidentate coordination

Modes	Observed[a]	Calculated[b]	SQM[c]	PED (≥10 %)
A symmetry				
1	1329 s	1387	1394	v_s S=O (51) + v_s SO$_2$ ip (30) + v_a SO$_2$ op (10)
2	1178 sh	1127	1179	v_a SO$_2$ op (46) + v_s S=O (38)
3	1055 sh	1109	1010	v_s SO$_2$ ip (48) + v_a SO$_2$ op (34)
4	1026 s	931	961	v_s Cr=O (95)
5	794 m	826	850	v_a S–F (76)
6	666 sh	657	679	δ SO$_2$ op (42) + v Cr–O (23)
7	580 sh	557	562	ρ SO$_2$ op (55) δ O–S–F ip (12) + v Cr–O (12)
8	534 w	556	509	wag SO$_2$ op (61) + δ O–S–F ip (17)
9	476 sh	478	475	δO–S–F ip (27) + δCrO$_2$ (18) + vCr–O (14) + ρSO$_2$ op (14) + δSO$_2$ ip (11)
10		436	445	δCrO$_2$ (56) + δ O–S–F ip (19) + τw CrO$_2$ ip (13)
11		347	341	τSO$_2$ ip (80)
12		270	274	δ SO$_2$ ip (21) + vCr–O (10) + δ O–S–F ip (18) +ρCrO$_2$ op (12) + τ CrO$_2$ ip (11)
13		235	220	ρ CrO$_2$ (64) + τ CrO$_2$ ip (16)
14		155	148	v Cr–O (32) + τ CrO$_2$ ip (20) + δ Cr–O$_2$ ip (11)
15		143	136	v_s Cr–O (41)
16		98	93	ρ CrO$_2$ (36) + δ Cr–O$_2$ ip (35) + τSO$_2$ ip (12) + τw CrO$_2$ ip (11)
17		57	56	τ Cr–O–S–O ip (75) + δ Cr–O$_2$ ip (16)
B symmetry				
18	1233 sh	1374	1379	v_a S=O (52) + v_s SO$_2$ ip (30) + v_a SO$_2$ op (12)
19	1209 vs	1149	1191	v_a SO$_2$ ip (42) + v_a S=O (40)
20	1026 s	1091	979	v_s SO$_2$ op (47) + v_a SO$_2$ ip (27) + v_a Cr=O (14)
21	989 sh	895	946	v_a Cr=O (95)
22	871 w	826	853	v_s S–F (72)
23	734 m	657	684	δSO$_2$ ip (64) + v Cr–O (24)
24	611 sh	557	578	ρ SO$_2$ ip (39) + v Cr–O (18) + δ O–S–F op (15) + δSO$_2$ op (12)
25	514 w	506	506	wag SO$_2$ ip (59) + δ O–S–F op (17) + δSO$_2$ op (10)
26	450[d]	449	462	δO–S–F op (38) + δSO$_2$ op (18) + ρSO$_2$ ip (17) + δSO$_2$ op (11) + vCr–O (10)
27		361	355	wag CrO$_2$ (33) + δ O–S–F op (17) + ρ SO$_2$ ip (16) + vCr–O (14) + δSO$_2$ op (10)
28	330[d]	336	334	τSO$_2$ op (76) + δCrO$_2$ op (10)
29		257	240	wag CrO$_2$ (57)
30		214	204	τw CrO$_2$ (63) + δCrO$_2$ op (12)
31		132	128	v Cr–O (32) + δCrO$_2$ op (12)
32		75	74	τ Cr–O–S–O op (73) + δCrO$_2$ op (18)
33		10	10	v Cr–O (38) + τw CrO$_2$ op (15)

[a] This work
[b] DFT B3LYP/6-31G*
[c] From scaled quantum mechanics force field
[d] From Ref. [23]

with a PED contribution of 72 % by using 6-31G* basis set [12]. The four Cr–O symmetric stretching modes are calculated at 355, 274, 136, and 10 cm^{-1} and assigned according to the calculations. The last two bands are attributed to the Cr ← O antisymmetric and symmetric stretchings. Also, the two modes, CrO$_2$ in-phase and out-of-phase bending (O–Cr–O), are calculated at 148 and 128 cm^{-1}, respectively, and for this both modes were not assigned. The wagging, rocking, and twisting modes of the CrO$_2$ group are calculated strongly mixed with other modes. The first two modes are assigned as in the monodentate case (see Table 2.8), while the out-of-phase and in-phase twisting modes are calculated and assigned at 204 and 93 cm^{-1}, respectively.

In this compound, both structures are probably present in the solid phase because the comparison of each infrared spectrum (Fig. 2.3a, b) with the corresponding experimental ones is different between them, whereas a comparison between the average calculated infrared spectra, as explained before, demonstrates good correlation (Fig. 2.3c).

The average theoretical spectrum is as displaced in relation to the experimental ones, as shown in Fig. 2.4. In addition, the proposed theoretical Raman spectra for both structures of the compound are presented in Fig. 2.5.

2.6 Force Field

The harmonic force field and the force constants for chromyl fluorosulfate were calculated by using the scaling procedure of Pulay et al. [25, 26] with the

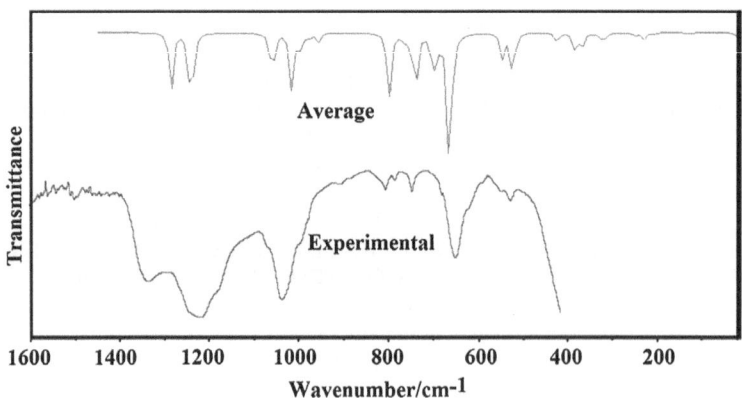

Fig. 2.4 Comparison between the infrared experimental spectrum of chromyl fluorosulfate with the calculated average infrared spectra for both structures from B3LYP/6–31G* wavenumbers and intensities using Lorentzian band shapes (for a population relation C$_2$(1): C$_2$(2) of 1: 1 for each structure). Reprinted from Journal of Molecular Structure, 981/1–3, A. Ben Altabef, S.A. Brandán, a new vibrational study of chromyl fluorosulfate, CrO2 (SO3F)2 by DFT calculations 146–152, copyright 2010, with permission from Elsevier

Fig. 2.5 Raman spectra for the theoretical C$_2$(1) and C$_2$(2) structures of chromyl fluorosulfate at B3LYP/6–31G* theory level. Reprinted from Journal of Molecular Structure, 981/1–3, A. Ben Altabef, S.A. Brandán, a new vibrational study of chromyl fluorosulfate, CrO2 (SO3F)2 by DFT calculations 146–152, copyright 2010, with permission from Elsevier

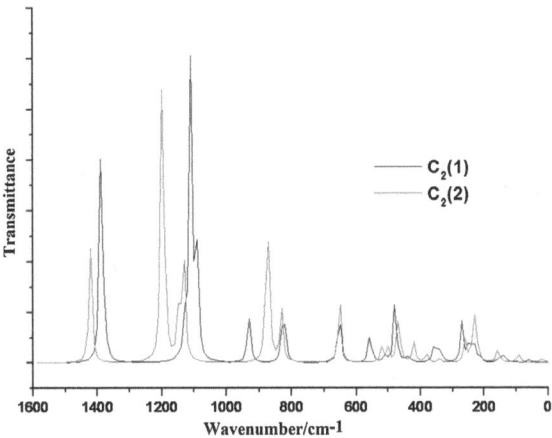

Table 2.12 Comparison of scaled internal force constants for chromyl fluorosulfate.

Force constant	B3LYP/6-31G*					
	[a]CrO$_2$(SO$_3$F)$_2$			[b]CrO$_2$(NO$_3$)$_2$		
	C$_2$(1)		C$_2$(2)	C$_2$		C$_{3V}$
	M	B	M	M	B	[a]SO$_3$F$^-$
f (S=O)	10.1	10.0	10.6			9.0
f (S–O)	6.4	8.6	4.7			
f (Cr=O)	6.6	6.6	8.8	6.55	6.53	
f (Cr–O)	2.8	1.3	3.7	6.09	1.44	
f (S–F)	5.2	5.2	4.8			4.0
f (O=S=O)	2.2	4.8	1.6			2.1
f (O=Cr=O)	2.4	2.3	2.1	2.53	1.66	
f (O–Cr–O)		0.8		0.80	0.93	
f (S–O–Cr)	1.8		0.4			
f (O–S–F)	2.0	2.6	1.7			1.6

Units are mdyn Å^{-1} for stretching and stretching/stretching interaction and mdyn Å rad^{-2} for angle deformations

Abbreviations: *M* monodentate, *B* bidentate

[a] This work

[b] Ref. [12]

Reprinted from Journal of Molecular Structure, 981/1–3, A. Ben Altabef, S.A. Brandán, a new vibrational study of chromyl fluorosulfate, CrO2 (SO3F)2 by DFT calculations 146–152, copyright 2010, with permission from Elsevier

MOLVIB program [51, 52], as mentioned before. The calculated forces constants for both structures appear in Table 2.12.

In general, the calculated force constants for a bidentate coordination of the fluorosulfate groups in the C$_2$(1) structure, with the B3LYP/6-31G* method, are approximately the same as the monodentate coordination of the same structure, with some variations in the f(S–O), f(Cr–O), f(O=S=O), and f(O−S−F) force constants. The S–O and Cr–O stretching force constants change with the coordination mode of the fluorosulfate group, being the first one greater in the bidentate coordination (8.6 mdyn Å^{-1}) than in the monodentate coordination (6.4 mdyn Å^{-1}) because, in this case, there are two S–O stretchings in each fluorsulfate group (Fig. 1), while the Cr–O stretching force constant is greater in the monodentate coordination (2.8 mdyn Å^{-1}), as it was also observed in $CrO_2(NO_3)_2$ [12], because in the bidentate coordination there are four Cr–O stretchings. Also, the above reasons justify that the f(O=S=O) is greater in the bidentate coordination than the other coordination mode. On the other hand, the differences between the force constant values in the monodentate coordination of both structures are attributed to the geometrical parameters. Thus, the lower values of the force constants of O=S=O deformations in the $C_2(2)$ structure of chromyl fluorosulfate can be attributed to the higher values of the O=S=O angles (109.5°) in this structure than in the other ones (106.7°), while the lower f(O–S–F) force constant values in the $C_2(2)$ structure (1.7 mdyn Å rad^{-2}) are associated to the O–S–F angles values, because they are slightly higher (107.1°) in this structure than the corresponding values of the $C_2(1)$ structure (107.8°). The S=O stretchings force constants considering the two structures and coordination modes are near to the expected values reported by other compounds that contain SO_2 groups [49, 50]. The analysis of the force constants of the compound with the values for $CrO_2(NO_3)_2$ suggests that the coordination that better represents the fluorosulfate group in chromyl fluorosulfate is the bidentate ($C_2(1)$ structure), while when those values are compared with the SO_3F^- ion, the coordination that better represents the fluorosulfate group in chromyl fluorosulfate is the monodentate ($C_2(2)$ structure). This way, both coordination modes are possible in chromyl fluorosulfate.

2.7 Conclusions

The chromyl fluorosulfate molecule was synthesized and characterized by infrared spectroscopic techniques in the Nujol suspension.

The presence of both coordination modes was detected in the IR spectrum, and a complete assignment of the vibrational modes were accomplished.

The calculations suggest the existence of two molecular $C_2(1)$ and $C_2(2)$ structures for chromyl fluorosulfate, both of C_2 symmetries, which probably origin the different colorations observed by other authors in different preparations of the compound.

An SQM/B3LYP/6-31G* force field was obtained for both structures of chromyl fluorosulfate after adjusting the obtained theoretical force constants to minimize the difference between the observed and calculated wavenumbers.

The NBO and AIM analysis confirm the hexacoordination of the Cr atom in chromyl fluorosulfate.

Acknowledgments This work was subsidized with grants from CIUNT (Consejo de Investigaciones, Universidad Nacional de Tucumán), and CONICET (Consejo Nacional de Investigaciones Científicas y Técnicas, R. Argentina). The authors thank Prof. Tom Sundius for his permission to use MOLVIB.

References

1. E.L. Varetti, S.A. Brandán, A. Ben Altabef, Vib. Spectros. **5**, 219 (1993)
2. S.A. Brandán, A. Ben Altabef, E.L. Varetti, Spectrochim. Acta **51A**, 669 (1995)
3. S.A. Brandán, A. Ben Altabef, E.L. Varetti, J. Raman Spectrosc. **27**, 447 (1996)
4. S.A. Brandán, Estudio Espectroscópico de Compuestos Inorgánicos Derivados de Metales de Transición, Doctoral Thesis, National University of Tucumán, R. Argentina, 1997
5. S.A. Brandán, A. Ben Altabef, E.L. Varetti, Anales de la Asoc.Qca. Arg. **87**(1,2), 89 (1999)
6. O.E. Piro, E.L. Varetti, S.A. Brandán, A. Ben Altabef, J. Chem. Cryst. **33**, 57 (2003)
7. M. Fernández Gómez, A. Navarro, S.A. Brandán, C. Socolsky, A. Ben Altabef, E.L. Varetti, J. Mol. Struct. (Theochem) **626**, 101 (2003)
8. C. Socolsky, S.A. Brandán, A. Ben Altabef, E.L. Varetti, J. Mol Struct. (Theochem) **672**, 45 (2004)
9. M.L. Roldán, H. Lanús, S.A. Brandán, J.J. López, E.L. Varetti, A. Ben Altabef, J. Argent. Chem. Soc. **92**, 53 (2004)
10. M.L. Roldán, S.A. Brandán, E.L. Varetti, A. Ben Altabef, Z. Anorg, Allg. Chem. **632**, 2495 (2006)
11. S.A. Brandán, M.L. Roldán, C. Socolsky, A. Ben Altabef, Spectrochim. Acta, Part A **69**, 1027 (2008)
12. S.A. Brandán, C. Socolsky, A. Ben Altabef, Z. Anorg. Allg. Chem. **635**(3), 582 (2009)
13. S.A. Brandán, J. Mol. Struc. (Theochem) **908**, 19 (2009)
14. C.C. Addison, N. Logan, S.C. Wallwork, C.D. Garner, Q. Rev. Chem. **25**, 289 (1971)
15. C.C. Addison, N. Logan, Adv. Inorg. Chem. Radiochim. **6**, 71 (1964)
16. C.C. Addison, D. Sutton, Prog. Inorg. Chem. **8**, 195 (1967)
17. W.A. Guillory, M.L. Bernstein, J. Chem. Phys. **62**(3), 1059 (1975)
18. J. Laane, J.R. Ohlsen, Prog. Inorg. Chem. **27**, 465 (1980)
19. B. Lippert, C.J.L. Lock, B. Rosenberg, M. Zvagulis, Inorg. Chem. **16**, 1525 (1977)
20. C.J. Marsden, K. Hedberg, M.M. Ludwig, G.L. Gard, Inorg. Chem. **30**, 4761 (1991)
21. M. Lustig, G.H. Cady, Inorg. Chem. **1**(3), 714 (1962)
22. W.V. Rochat, G.L. Gard, Inorg. Chem. **1**(8), 158 (1969)
23. S.D. Brown, G.L. Gard, Inorg. Chem. **14**(9), 2273 (1975)
24. S. Bell, T.J. Dines, J. Phys. Chem. A **104**, 11403 (2000)
25. G. Rauhut, P. Pulay, J. Phys. Chem. **99**, 3093 (1995)
26. G. Rauhut, P. Pulay, J. Phys. Chem. **99**, 14572 (1995)
27. F. Kalincsák, G. Pongor, Spectrochim. Acta A **58**, 999 (2002)
28. A.E. Reed, L.A. Curtis, F. Weinhold, Chem. Rev. **88**(6), 899 (1988)
29. J.P. Foster, F. Weinhold, J. Am. Chem. Soc. **102**, 7211 (1980)
30. A.E. Reed, F. Weinhold, J. Chem. Phys. **83**, 1736 (1985)
31. E.D. Gledening, J.K. Badenhoop, A.D. Reed, J.E. Carpenter, F. Weinhold, *NBO 3.1* (Theoretical Chemistry Institute, University of Wisconsin, Madison, 1996)
32. R.F.W. Bader, *Atoms in Molecules, A Quantum Theory* (Oxford University Press, Oxford, 1990). ISBN: 0198558651
33. F. Biegler-Köning, J. Schönbohm, D. Bayles, AIM2000; a program to analyze and visualize atoms in molecules. J. Comput. Chem. **22**, 545 (2001)
34. A.B. Nielsen, A.J. Holder, *GaussView* (User's Reference, Gaussian, Inc, Pittsburgh, 2000–2003)

35. M.J. Frisch, J.A. Pople, *Gaussian 03, Revision B.01* (Gaussian, Inc., Pittsburgh, 2003)
36. A.D. Becke, J. Chem. Phys. **98**, 5648 (1993)
37. C. Lee, W. Yang, R.G. Parr, Phys. Rev. B **37**, 785 (1988)
38. P.J. Perdew, Phys. Rev. B **33**, 8822 (1986)
39. P.J. Perdew, K. Burke, Y. Wang, Phys. Rev. B **54**, 16533 (1996)
40. A.W. Ross, M. Fink, R. Hilderbrandt, in *International Tables for Crystallography*, vol. C ed. by A.J.C. Wilson (Kluwer Academic Publishers, Dordrecht, Boston and London, p. 245, 404 1992)
41. Z. Žak, M. Kosička, Acta Cryst. **B34**, 38 (1978)
42. R.J. Gillespie (ed.), *Molecular Geometry* (Van Nostrand-Reinhold, London, 1972)
43. R.J. Gillespie, I. Bytheway, T.H. Tang, R.F.W. Bader, Inorg. Chem. **35**, 3954 (1996)
44. G.L. Sosa, N. Peruchena, R.H. Contreras, E.A. Castro, J. Mol. Struct. (Theochem) **401**, 77 409 (1997)
45. J. Goubeau, J.B. Milne, Canad. J. Chem. **45**, 2321 (1967)
46. R.J. Gillespie, R.A. Rothenbury, Can. J. Chem. **42**, 416 (1964)
47. R.J. Gillespie, E.A. Robinson, Can. J. Chem. **40**, 644 (1962)
48. A. Goypiron, J. De Villepin, A. Novak, J. Chim. Phys. **76**(3), 267 (1979)
49. H. Siebert, *Anwendungen der schwingungsspektroskopie in der Anorganische Chemie* (Springer, Berlin, 1966), p. 72
50. K. Nakamoto, *Infrared and Raman Spectra of Inorganic and Coordination Compounds*, 5th edn. (Wiley, New York 1997)
51. T. Sundius, J. Mol. Struct. **218**, 321 (1990)
52. T. Sundius, *MOLVIB: A Program for Harmonic Force Field Calculation*, QCPE Program No. 604 (1991)

Chapter 3
Structural and Vibrational Study of Chromyl Nitrate

Abstract The density functional theory (DFT) has been used to study the structural and vibrational properties of chromyl nitrate. The molecular structure of the compound has been theoretically determined in gas phase employing the B3LYP, B3P86, and B3PW91 levels of theory, and the harmonic vibrational frequencies were evaluated at the same levels. The calculated harmonic vibrational frequencies for chromyl fluorosulphate are consistent with the experimental IR spectrum. These calculations gave us a precise knowledge of the normal modes of vibration taking into account the type of coordination adopted by fluorosulphate groups of this compound as monodentate and bidentate ligands. Also, the assignment of all the observed bands in the IR spectrum for chromyl fluorosulphate was performed. The nature of the Cr–O and Cr ← O bonds and the topological properties of the compound were investigated and analyzed by means of natural bond order (NBO) and *Bader's* atoms in molecules theory (AIM), respectively.

Keywords Chromyl chlorosulfate · Vibrational spectra · Molecular structure · Force field · DFT calculations

3.1 Introduction

The study of compounds that contain transition metals, such as V and Cr [1–6] is of great interest; especially those with the nitrate group as ligand because this group is a versatile ligand and can act as monodentate or bidentate ligand [7]. The

Adapted from Spectrochimica Acta part A: Molecular and Biomolecular Spectroscopy, 69/3, S.A. Brandan, M.L. Roldan, C. Socolsky, A. Ben Altabef, 1027-1043, Copyright 2008, with permission from Elsevier

S. A. Brandán, *A Structural and Vibrational Study of the Chromyl Chlorosulfate, Fluorosulfate, and Nitrate Compounds*, SpringerBriefs in Molecular Science, DOI: 10.1007/978-94-007-5763-9_3, © The Author(s) 2013

mode of coordination adopted by nitrate groups and the stereochemistry of this compound are important in relation to the vibrational properties and chemical reactivity [7–9]. The chromyl nitrate, $CrO_2(NO_3)_2$, compound presents vibrational properties imperfectly described and only the main characteristics of the infrared (IR) spectrum were published in the liquid phase [10] and in previous studies we have assigned some bands observed in the vibrational spectra of the chromyl nitrate [1]. Marsden et al. have studied the gas-phase molecular structure of chromyl nitrate, $CrO_2(NO_3)_2$, by electron diffraction at a temperature of 50 °C and by ab initio methods at the HF level [11]. In this chapter, the diffraction data are consistent with C_2 symmetry for the molecule. The Cr coordination is best described as derived from a severely distorted octahedron, since the nitrate groups act as bidentate ligands which are asymmetrically bonded to Cr. This dark red liquid compound can be obtained by several methods [10–15] and it is very reactive at room temperature but less unstable than other compounds with the chromyl group [16, 17]. In this chapter, an experimental and theoretical study on this compound with the methods of quantum chemistry was carried out in order to have a better understanding of its vibrational properties. A precise knowledge of the normal modes of vibration is expected to provide a foundation for understanding the conformation-sensitive bands in vibrational spectra of this molecule. In this study, the normal mode calculations were accomplished using a generalized valence force field (GVFF) and considering the nitrate group as monodentate and bidentate ligand. In this study, an experimental and theoretical study of chromyl nitrate, $CrO_2(NO_3)_2$ was performed, in order to study the coordination mode of nitrate groups and carry out its complete assignment. For this purpose, the optimized geometry and frequencies for the normal modes of vibration were calculated. In this case, there are no publications about experimental or high-level theoretical studies on the geometries and force field of chromyl nitrate. Hence, obtaining reliable parameters by theoretical methods is an appealing alternative. The parameters obtained may be used to gain chemical and vibrational insights into related compounds.

   The election of the method and the basis sets are very important to evaluate not only the best level of theory but also the best basis set to be used to reproduce the experimental geometry and the vibrational frequencies. In previous studies of compounds that contain metal transition such as $VO_2X_2^-$ (X = F, Cl) anions [3], the HF and MP2 methodologies are much less satisfactory than the DFT techniques especially for the V–Cl distance. In this case, the basis set that best reproduces the experimental geometric parameters for the chloro compound is B3PW91/6-311G* while the inclusion of polarization functions is important to have a better agreement. In the VOX_3 (X = F, Cl, Br, I) series, the optimized geometry which better reproduces the experimental parameters was obtained with the B3PW91/6-311G calculation while the B3LYP method produces the best results for the vibrational frequencies [4]. In a recent paper about oxotetrachlorochromate (V) anion [6], it was found that the inclusion of polarization functions in the basis sets significantly improved the theoretical geometry results and the lowest deviation with reference to the experimental data was obtained for the 6-31G* and

6-311G* basis sets and the B3PW91 functional [6]. In this case, the lower difference between theoretical and experimental frequencies, measured by the root means standard deviation (RMSD) was obtained with the combination B3LYP/6-31 + G. Similar results were obtained in the study of the $[VOCl_4]^-$ anion [5]. In the study of the structures and vibrational spectra of chromium oxoanions and oxyhalide compounds, Bell et al. [18] have found that B3LYP/Lanl2DZ combination gives the best fit for the geometries and observed vibrational spectra.

In this case, DFT calculations were used to study the structure and the vibrational properties of the compound. The normal mode calculations were accomplished by use of a GVFF. The results show that a molecular force field for the chromyl nitrate, considering the nitrate group as well as monodentate and bidentate ligand calculated using the DFT/Lanl2DZ, 6-31G*, and 6-311+G combinations, is well represented. The force field scaling factors used produce satisfactory agreement between the calculated and experimental vibrational frequencies of chromyl nitrate. DFT normal mode assignments, in terms of the potential energy distribution, are in general accord with those obtained from the normal coordinate analysis. Also, the nature of the two types of Cr–O and Cr ← O bonds in chromyl nitrate was systematically and quantitatively investigated by the NBO analysis [19]. In addition, the topological properties of electronic charge density are analyzed employing Bader's Atoms in Molecules theory (AIM) [19–30].

3.2 Structural Analysis

The numbering of the atoms for monodentate and bidentate chromyl nitrate is described in Fig. 3.1.

The B3LYP structures obtained from chromyl nitrate with the different basis sets have C_2 symmetries similar to the experimental structure obtained by diffraction data and HF method by Marsden et al. [11]. Table 3.1 shows the comparison of the total energies and dipole moments values for chromyl nitrate with the B3LYP method using different basis sets. In all cases, the more stable structure is obtained using the B3LYP/6-311+G* method combined with a diffuse function basis set while the structure with higher energy is obtained by B3LYP/Lanl2DZ calculation. The higher dipole moment value agrees with the one obtained with the 6-311+G* basis set and this indicates that the largest dipole moment value stabilizes the molecule. Although the calculated structure with the Lanl2DZ basis set is unstable, the dipole moment value is comparable to the corresponding values obtained with the 6-31G* and 6-31+G basis sets.

The results of the calculations with all basis sets used can be appreciated in Table 3.2. According to these results, the method and basis set that best reproduces the experimental geometric parameters for the chromyl nitrate compound is B3LYP/6-31+G* where the means difference for bond lengths is 0.019 Å, while with B3LYP/6-311G* it is 4.11° for angles. The inclusion of polarization functions, however, is important to have a better agreement with the experimental

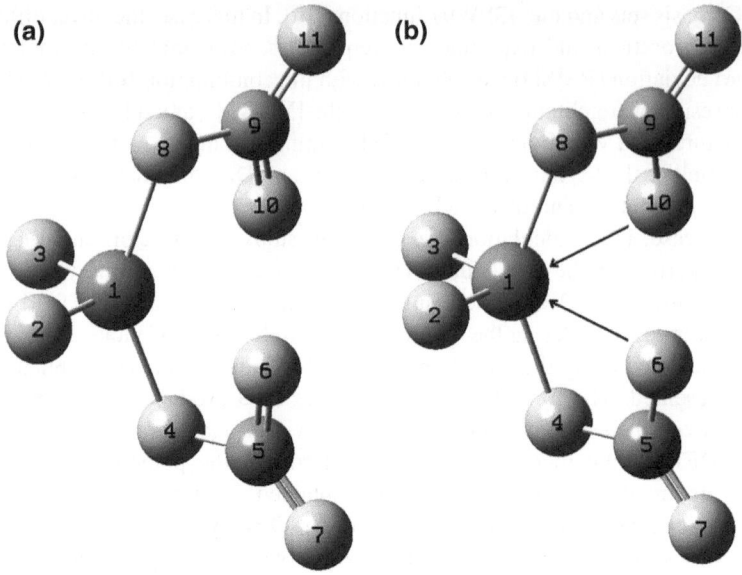

Fig. 3.1 The C$_2$(1) molecular structure of chromyl nitrate considering the nitrate group as: **a** monodentate ligand and **b** bidentate ligand. Reprinted from, [47], Copyright 2008, with permission from Elsevier

Table 3.1 Total energy and dipole moment for chromyl nitrate at B3LYP method

Basis set	ET (Hartree)	μ (D)
Lanl2DZ	−797.26788972	0.54
STO-3G	−1735.04954605	0.14
3-21G*	−1746.64837014	0.28
6-31G	−1755.21237919	0.43
6-31G*	−1755.48635040	0.50
6-311G	−1755.49143234	0.67
6-31+G	−1755.25752525	0.59
6-31+G*	−1755.51881813	0.71
6-311G*	−1755.73534229	0.70
6-311+G	−1755.52645780	0.63
6-311+G*	−1755.76567840	0.73

Reprinted from [47], Copyright 2008, with permission from Elsevier

geometry: means differences degrade to 0.020 Å and 4.24° for the 6-311+G* basis set. The B3LYP functional gives somewhat less satisfactory agreement using the Lanl2DZ (0.399 Å and 7.90°) and STO-3G (0.107 Å and 5.42) basis sets, contrary to what was observed by Bell et al. [18] in chromium oxoanions and oxyhalides compounds. In this case, similarly to the experimental structure [11], also with B3LYP calculations, it is possible to represent the coordination around Cr as

Table 3.2 Comparison of experimental and calculated geometric parameters at different levels of theory for chromyl nitrate.

Atoms	[a]B3LYP Method											[b]Exp.	[b]ab initio
	Lanl2DZ	STO-3G	3-21G	6-31G	6-31+G	6-31+G*	6-31G*	6-311G	6-311G*	6-311+G	6-311+G*		
Bond length (Å)													
1,2	1.568	1.484	1.561	1.573	1.577	1.554	1.548	1.569	1.546	1.574	1.550	1.586	1.522
1,3	1.568	1.484	1.561	1.573	1.577	1.554	1.548	1.569	1.546	1.574	1.550	1.586	1.522
1,4	1.914.	1.864	1.888	1.933.	1.935	1.923	1.920	1.933	1.916	1.941	1.924	1.957	1.958
1,8	1.914	1.864	1.888	1.933	1.935	1.923	1.920	1.933	1.916	1.941	1.924	1.957	1.958
1,6	1.409	2.155	2.205	2.275	2.287	2.243	2.229	2.283	2.253	2.284	2.263	2.254	2.191
1,10	1.310	2.155	2.205	2.275	2.287	2.243	2.229	2.283	2.253	2.284	2.263	2.254	2.191
4,5	1.231	1.436	1.432	1.397	1.396	1.340	1.340	1.394	1.342	1.393	1.342	1.341	1.342
5,6	1.409	1.372	1.325	1.296	1.296	1.261	1.261	1.298	1.255	1.298	1.254	1.254	1.342
5,7	1.310	1.294	1.222	1.216	1.218	1.195	1.194	1.219	1.185	1.220	1.186	1.193	1.205
8,9	1.231	1.436	1.432	1.397	1.396	1.340	1.340	1.394	1.342	1.393	1.342	1.341	1.342
9,10	1.568	1.372	1.325	1.296	1.296	1.261	1.261	1.298	1.257	1.298	1.254	1.254	1.342
9,11	1.568	1.294	1.222	1.216	1.218	1.195	1.194	1.219	1.185	1.220	1.186	1.193	1.205
RMSD (Å)	0.399	0.107	0.060	0.033	0.034	0.019	0.024	0.034	0.024	0.033	0.020		
Bond angle (°)													
2,1,3	108.2	107.5	107.1	107.8	107.9	108.0	107.9	108.0	108.0	107.9	108.1	112.6	105.0
2,1,4	105.2	103.2	104.6	104.9	105.1	105.5	105.2	104.8	104.8	105.2	105.4	97.2	105.5
2,1,8	97.3	96.2	96.9	97.3	97.3	97.0	97.1	96.9	97.5	97.1	97.5	104.5	105.5
3,1,4	97.3	96.2	96.9	97.3	97.3	97.0	97.1	96.9	97.5	97.1	97.5	104.5	105.5
3,1,8	105.2	103.2	104.6	104.9	105.1	105.5	105.2	104.8	104.8	105.2	105.4	97.2	105.5
4,1,8	141.2	146.9	143.4	141.9	141.6	141.3	141.6	142.6	141.6	141.6	140.6	140.4	146.9
6,1,10	101.0	72.6	73.2	73.5	73.9	75.0	74.5	74.3	74.6	73.8	74.5	82.8	
1,4,5	110.0	96.7	97.9	99.5	99.9	99.6	99.2	99.9	100.1	99.7	100.2	97.5	99.0
4,5,6	121.2	108.2	108.7	110.3	110.5	111.5	111.4	110.3	111.4	110.5	111.6	112.2	
4,5,7	128.7	123.3	122.1	121.3	121.2	120.9	120.9	121.3	120.8	121.3	120.7	119.7	

(continued)

Table 3.2 (continued)

	[a]B3LYP Method											[b]Exp.	[b]ab initio
	Lanl2DZ	STO-3G	3-21G	6-31G	6-31+G	6-31+G*	6-31G*	6-311G	6-311G*	6-311+G	6-311+G*		
6,5,7	101.0	128.4	129.2	128.4	128.3	127.6	127.6	128.4	127.8	128.2	127.7	128.1	
1,8,9	110.0	96.7	97.9	99.5	99.9	99.6	99.2	99.9	100.1	99.7	100.2	97.5	99.0
8,9,10	121.2	108.2	108.7	110.3	110.5	111.5	111.4	110.3	111.4	110.5	111.6	112.2	
8,9,11	128.7	123.3	122.1	121.4	121.2	120.9	120.9	121.3	120.8	121.3	120.7	119.7	
10,9,11	108.23	128.4	129.2	128.4	128.3	127.6	127.6	128.4	127.8	128.2	127.7	128.1	
RMSD (°)	7.90	5.42	4.49	4.23	4.27	4.36	5.00	4.33	4.11	4.33	4.24		
Bond dihedral angle (°)													
2,1,4,5	−82.6	−84.9	−84.2	−83.8	−83.2	−82.5	−83.0	−83.6	−83.1	−83.5	−82.8		
3,1,4,5	166.2	165.4	166.0	165.4	165.9	166.4	166.1	165.6	165.9	165.6	165.9		
8,1,4,5	40.6	39.4	39.8	39.7	40.2	40.7	40.3	39.9	40.4	39.9	40.4		
2,1,8,9	166.2	165.4	166.0	165.4	165.9	166.4	166.1	165.6	165.9	165.6	165.9		
3,1,8,9	−82.6	−84.9	−84.2	−83.8	−83.2	−82.5	−83.0	−83.6	−83.1	−83.5	−82.8		
4,1,8,9	40.6	39.4	39.8	39.7	40.2	40.7	40.3	39.9	40.4	39.9	40.4		
1,4,5,6	−1.3	−1.957	−1.6	−0.9	−1.3	−1.2	−1.1	−0.8	−1.1	−0.9	−0.9	16	
1,4,5,7	179.1	179.4	178.9	179.6	178.9	179.0	179.2	179.0	179.2	179.4	179.3		
1,8,9,10	−1.3	−195.7	−156.3	−0.9	−1.3	−1.2	−11.2	−0.8	−11.3	−0.9	−0.9		
1,8,9,11	179.1	179.4	178.9	179.6	178.9	179.0	179.2	179.7	179.2	179.4	179.3		

[a] This work
[b] Ref. [9]
Reprinted from [47], Copyright 2008, with permission from Elsevier

derived from a severely distorted octahedron where the nitrate groups act as bidentate ligands and are asymmetrically bonded to Cr. The bond orders, expressed by Wiberg's indexes for chromyl nitrate are given in Tables 3.3 and 3.4. Clearly, the chromium atom forms six bonds, two Cr=O bonds (bond order 1.9721 using 6-311+G basis set), two Cr–O (bond order 0.5374 using 6-311+G basis set), and two Cr ← O (bond order 0.1821 using 6-311+G basis set). The bond order value of this last bond was estimated by Marsden et al. [11] between 0.19 and 0.29. Experimentally, the Cr–O–NO$_2$ group is slightly nonplanar with the dihedral angle of the planes Cr–O–N and NO$_2$ equal to 16° while in our B3LYP calculations the Cr–O–N–O angle values are between 0.9 and 2.0°, as can be seen in Table 3.2. In this case, the two NO$_2$ groups are practically planar with the Cr atom and there is a slightly torsional motion around either of the Cr–O bonds as was observed by the diffraction method. However, the bond orders obtained for chromyl nitrate with the three basis sets agree with the 0.4 values reported in symmetrically bidentate nitrate groups [7].

Other very important observations in these B3LYP calculations carried out precisely by Marsden et al. [11] with ab initio method are those that predict that the O2=Cr1=O3 bond angle in chromyl nitrate is smaller than the O4–Cr1–O8 angle, in contradiction with the valence-shell electron-pair repulsion (VSEPR) theory [20, 21].

In fact, that theory predicts a larger angle between the bonds of the central atom with the two oxygen atoms in these pseudotetrahedral species as a consequence of the larger space in the coordination sphere requested by the pair of double bonds. In this case, a possible explanation for the inverse angle relationship is due to the octahedral coordination for the Cr atom in the compound. A possible explanation for the inverse relationship obtained for chromyl nitrate can be given in terms of the calculated Mulliken atomic charges. Such charges, taken from the DFT calculations (see Table 3.4 using 6-31G* and 6-311+G basis sets), have the following values: Cr 1.364, O2 −0.317, O3 −0.317, O4 −0.438, and O8 −0.438. These numbers show (using 6-31G* basis set) that the remarkably higher negative charge on the oxygen atoms of chromyl nitrate could result in a relatively larger O...O repulsion and consequently in a larger O4–Cr1–O8 angle (141.6°), which for such compound exceeds the O2–Cr1–O3 angle (107.9°). It is interesting to observe that these values of Mulliken atomic charges are very different from those obtained at B3LYP/Lanl2DZ and B3LYP/6-311+G* levels, only as regards the numerical values but not the relative order within them. The inclusion of diffuse functions makes the charge on the oxygen bound to the Cr atom to be lower (O=Cr=O) than that located on the other O atoms (O–Cr–O); in this case, these values are positive as shown in Table 3.4. This leads to the conclusion that any analysis based on the Mulliken atomic charges must be made with care as it is apparent that a satisfactory explanation based on electronegativity criteria is not affordable [3]. A somewhat different but related explanation might be tried on the basis of the delocalized and/or bonding characters of the relevant molecular orbitals (MO), as observed in the series of the VO$_2$X$_2^-$ anions [3]. After a careful inspection of the atomic orbital coefficients (AO) appearing in the different MO it is possible to note

Table 3.3 Wiberg Index bond matrix of chromyl nitrate at different levels of theory

B3LYP Method

Lanl2DZ Basis set

Atoms	1	2	3	4	5	6	7	8	9	10	11
1. Cr	0.0000	1.9633	1.9633	0.5830	0.0095	0.1781	0.0425	0.5830	0.0095	0.1781	0.0425
2. O	1.9633	0.0000	0.3192	0.1194	0.0056	0.0139	0.0135	0.0785	0.0123	0.0831	0.0091
3. O	1.9633	0.3192	0.0000	0.0785	0.0123	0.0831	0.0091	0.1194	0.0056	0.0139	0.0135
4. O	0.5830	0.1194	0.0785	0.0000	1.0070	0.1507	0.2018	0.0473	0.0022	0.0091	0.0039
5. N	0.0095	0.0056	0.0123	1.0070	0.0000	1.3219	1.6331	0.0022	0.0006	0.0024	0.0009
6. O	0.1781	0.0139	0.0831	0.1507	1.3219	0.0000	0.3379	0.0091	0.0024	0.0027	0.0028
7. O	0.0425	0.0135	0.0091	0.2018	1.6331	0.3379	0.0000	0.0039	0.0009	0.0028	0.0019
8. O	0.5830	0.0785	0.1194	0.0473	0.0022	0.0091	0.0039	0.0000	1.0070	0.1507	0.2018
9. N	0.0095	0.0123	0.0056	0.0022	0.0006	0.0024	0.0009	1.0070	0.0000	1.3219	1.6331
10. O	0.1781	0.0831	0.0139	0.0091	0.0024	0.0027	0.0028	0.1507	13.219	0.0000	0.3379
11. O	0.0425	0.0091	0.0135	0.0039	0.0009	0.0028	0.0019	0.2018	16.331	0.3379	0.0000
6-31G* Basis set											
1. Cr	0.0000	1.9623	1.9623	0.4975	0.0060	0.1713	0.0436	0.4975	0.0060	0.1713	0.0436
2. O	1.9623	0.0000	0.2364	0.0723	0.0046	0.0114	0.0106	0.0633	0.0058	0.0579	0.0083
3. O	1.9623	0.2364	0.0000	0.0633	0.0058	0.0579	0.0083	0.0723	0.0046	0.0114	0.0106
4. O	0.4975	0.0723	0.0633	0.0000	1.0795	0.1412	0.2018	0.0234	0.0011	0.0066	0.0030
5. N	0.0060	0.0046	0.0058	1.0795	0.0000	1.3375	1.6314	0.0011	0.0003	0.0017	0.0008
6. O	0.1713	0.0114	0.0579	0.1412	1.3375	0.0000	0.3107	0.0066	0.0017	0.0033	0.0025
7. O	0.0436	0.0106	0.0083	0.2018	1.6314	0.3107	0.0000	0.0030	0.0008	0.0025	0.0016
8. O	0.4975	0.0633	0.0723	0.0234	0.0011	0.0066	0.0030	0.0000	1.0795	0.1412	0.2018
9. N	0.0060	0.0058	0.0046	0.0011	0.0003	0.0017	0.0008	1.0795	0.0000	1.3375	1.6314
10. O	0.1713	0.0579	0.0114	0.0066	0.0017	0.0033	0.0025	0.1412	1.3375	0.0000	0.3107
11. O	0.0436	0.0083	0.0106	0.0030	0.0008	0.0025	0.0016	0.2018	1.6314	0.3107	0.0000
6-311+G Basis set											

(continued)

Table 3.3 (continued)

B3LYP Method

Lanl2DZ Basis set

Atoms	1	2	3	4	5	6	7	8	9	10	11
1. Cr	0.0000	1.9721	1.9721	0.5374	0.0072	0.1821	0.0446	0.5374	0.0072	0.1821	0.0446
2. O	1.9721	0.0000	0.2908	0.0946	0.0066	0.0126	0.0127	0.0654	0.0096	0.0721	0.0091
3. O	1.9721	0.2908	0.0000	0.0654	0.0096	0.0721	0.0091	0.0946	0.0066	0.0126	0.0127
4. O	0.5374	0.0946	0.0654	0.0000	1.0329	0.1521	0.2050	0.0363	0.0017	0.0070	0.0037
5. N	0.0072	0.0066	0.0096	1.0329	0.0000	1.3196	1.6222	0.0017	0.1329	0.0026	0.0011
6. O	0.1821	0.0126	0.0721	0.1521	1.3196	0.0000	0.3265	0.0070	0.0026	0.0039	0.0032
7. O	0.0446	0.0127	0.0091	0.2050	1.6222	0.3265	0.0000	0.0037	0.0011	0.0032	0.0019
8. O	0.5374	0.0654	0.0946	0.0363	0.0017	0.0070	0.0037	0.0000	1.0329	0.1521	0.2050
9. N	0.0072	0.0096	0.0066	0.0017	0.0004	0.0026	0.0011	1.0329	0.0000	1.3196	1.6222
10. O	0.1821	0.0721	0.0126	0.0070	0.0026	0.0039	0.0032	0.1521	1.3196	0.0000	0.3265
11. O	0.0446	0.0091	0.0127	0.0037	0.0011	0.0032	0.0019	0.2050	1.6222	0.3265	0.0000

Table 3.4 Wiberg Index and atomic charges of chromyl nitrate at different level of theory

	Lanl2DZ		6-31G*		6-311+G	
	Atomic charges	Wiberg index	Atomic charges	Wiberg index	Atomic charges	Wiberg index
1. Cr	0.756579	5.5527	1.363988	5.3616	0.652634	5.4868
2. O	−0.151948	2.6179	−0.316872	2.4328	−0.208349	2.5456
3. O	−0.151948	2.6179	−0.316872	2.4328	−0.208349	2.5456
4. O	−0.261627	2.2028	−0.438030	2.0898	0.257663	2.1359
5. N	0.290469	3.9955	0.792440	4.0688	−0.720529	4.0038
6. O	−0.174453	2.1026	−0.413169	2.0442	0.219369	2.0815
7. O	−0.080730	2.2475	−0.306362	2.2143	0.125529	2.2298
8. O	−0.261627	2.2028	−0.438030	2.0898	0.257663	2.1359
9. N	0.290469	3.9955	0.792440	4.0688	−0.720529	4.0038
10. O	−0.174453	2.1026	−0.413169	2.0442	0.219369	2.0815
11. O	−0.080730	2.2475	−0.306362	2.2143	0.125529	2.2298

that, as a general pattern, the highest occupied MOs show a rather localized character on the oxygen atoms being mainly described as p-type orbitals. The atomic orbital AO coefficients for Cr atom of chromyl nitrate (*d*-type orbitals) using Lanl2DZ, 6-31G*, and 6-311+G basis sets are observed in Table 3.5. For chromyl nitrate, the strongest bonding MOs involving Cr *d*-type orbitals that seem to be sensitive to the geometry can be considered, in increasing energy, those numbered as 13 (HOMO-31), 13 (HOMO-36), and 14 (HOMO-32) calculated with Lanl2DZ basis set; 20 (HOMO-36), 21 (HOMO-36), and 22 (HOMO-37) calculated with 6-31G* basis set while those numbered as 25 (HOMO-36), 26 (HOMO-37), and 31 (HOMO-37) calculated with 6-311+G basis set tend to widen the O4–Cr1–O8 angle (125.3° using 6-31G* basis set) on a maximum overlapping basis.

For chromyl nitrate, the intermolecular interactions have been analyzed by using Bader's topological analysis of the charge electron density, $\rho(r)$ by means of the AIM program [23]. It is necessary to clarify that in this study, the Lanl2DZ, 6-31G*, and 6-311+G basis sets have been considered because there are numerous references where the quality of the basis set has no influence on the topological results [24, 25]; but, in this case, there is a significant difference among them as can be seen in Table 3.6.

The localization of the critical points in the $\rho(r)$ and the values of the Laplacian at these points are important for the characterization of molecular electronic structure in terms of interactions' nature and magnitude.

The details of the molecular models for the compound studied showing the geometry of all their critical points are observed in Fig. 3.2.

The analyses of the Cr ← O bond critical points in the compound studied are reported with the two basis sets in Table 3.6. In this case, two important observations can be seen for the three basis sets. In one case, the Cr1 ← O6 and Cr1 ← O10 bond critical points have the typical properties of the closed-shell interaction. That is, the values of $\rho(r)$ are relatively low (0.05 and 0.3 a.u.), the

Table 3.5 Atomic orbital coefficients (AO) for Cr atom of chromyl nitrate at different levels of theory B3LYP Method

Lanl2DZ				6-31G*				6-311+G			
N° orbital	Type orbital	HOMO-31	HOMO-32	N° orbital	Type orbital	HOMO-36	HOMO-37	N° orbital	Type orbital	HOMO-36	HOMO-37
13	7D 0	0.33926	0.00000	18	6XX	-0.33178	0.00000	25	15D 0	0.15892	0.00000
14	7D + 1	0.00000	0.46144	19	6YY	0.07504	0.00000	26	15D + 1	0.00000	0.21345
15	7D - 1	0.00000	-0.24884	20	6ZZ	0.24774	0.00000	27	15D - 1	0.00000	-0.11802
16	7D + 2	-0.28376	0.00000	21	6XY	0.18994	0.00000	28	15D + 2	-0.13523	0.00000
17	7D - 2	0.24434	0.00000	22	6XZ	0.00000	0.35911	29	15D - 2	0.11457	0.00000
18	8D 0	0.14192	0.00000	23	6YZ	0.00000	-0.19411	30	16D 0	0.16165	0.00000
19	8D + 1	0.00000	0.11151	24	7XX	-0.20429	0.00000	31	16D + 1	0.00000	0.22423
20	8D - 1	0.00000	-0.06622	25	7YY	0.04488	0.00000	32	16D - 1	0.00000	-0.12192
21	8D + 2	-0.09657	0.00000	26	7ZZ	0.16284	0.00000	33	16D + 2	-0.14011	0.00000
22	8D - 2	0.07986	0.00000	27	7XY	0.11686	0.00000	34	16D - 2	0.11908	0.00000
				28	7XZ	0.00000	0.20458	35	17D 0	0.15837	0.20227
				29	7YZ	0.00000	-0.11549	36	17D + 1	0.00000	-0.11466
				30	8F0	0.01138	0.00000	37	17D - 1	0.00000	0.00000
				31	8F + 1	0.00000	0.00858	38	17D + 2	-0.12927	0.00000
				32	8F - 1	0.00000	-0.00403	39	17D - 2	0.10925	0.00000
				33	8F + 2	0.00416	0.00000	40	18S	-0.00120	0.00000
				34	8F - 2	0.00292	0.00000	41	19PX	0.00000	0.02540
				35	8F + 3	0.00000	-0.00043	42	19PY	0.00000	0.04889
				36	8F - 3	0.00000	-0.00425	43	19PZ	0.06290	0.00000
								44	20PX	0.00000	-0.00099
								45	20PY	0.00000	0.00879
								46	20PZ	0.04372	0.00000
								47	21D 0	0.05593	0.00000
								48	21D + 1	0.00000	0.00350
								49	21D - 1	0.00000	-0.00900
								50	21D + 2	0.01428	0.00000
								51	21D - 2	-0.01112	0.00000

Table 3.6 Analysis of Cr ← O bond critical points in chromyl nitrate. (The quantities are in atomics units)

Bond	B3LYP Method Lanl2DZ					
	Cr1 ← O6	Cr1 ← O6	Cr1 ← O10	Cr1 ← O10	(3, +1)	(3, +1)
$\rho(r)$	0.364793	0.040763	0.364793	0.040763	0.032569	0.032569
$\nabla^2\rho(r)$	**−0.087669**	0.197670	**−0.087669**	0.197670	0.191651	0.191651
$\lambda 1$	−0.876120	−0.048900	−0.876120	−0.048900	−0.031081	−0.031081
$\lambda 2$	−0.800635	−0.047673	−0.800635	−0.047673	0.070519	0.070519
$\lambda 3$	1.589086	0.294234	1.589086	0.294234	0.152212	0.152212
$|\lambda 1|/\lambda 3$	0.551326	0.166192	0.551326	0.166192	0.204197	0.204197
B3LYP/6-31G*						
$\rho(r)$	0.4602388	0.04979	0.4602388	0.04979	0.040733	0.040733
$\nabla^2\rho(r)$	**−0.809380**	0.21576	**−0.809380**	0.21576	0.185820	0.185745
$\lambda 1$	−1.168065	−0.06250	−1.168065	−0.06250	−0.042819	−0.042811
$\lambda 2$	−1.012724	−0.05933	−1.012724	−0.05933	0.071014	0.071014
$\lambda 3$	1.371408	0.33760	1.371408	0.33760	0.157627	0.157533
$|\lambda 1|/\lambda 3$	0.851730	0.18513	0.851730	0.18513	0.271650	0.271758
B3LYP/6-311+G						
$\rho(r)$	0.379166	0.301963	0.379166	0.301963	0.0374596	0.0374596
$\nabla^2\rho(r)$	**−0.186240**	0.043910	**−0.186240**	0.043910	0.015960	0.015960
$\lambda 1$	−0.919694	−0.700007	−0.919694	−0.700007	−0.0342722	−0.0342722
$\lambda 2$	−0.852381	−0.641803	−0.852381	−0.641803	0.0578309	0.0578309
$\lambda 3$	1.585836	1.385788	1.585836	1.385788	0.1360750	0.1360750
$|\lambda 1|/\lambda 3$	0.57990	0.505200	0.57990	0.505200	0.250620	0.250620

relationship $|\lambda 1|/\lambda 3$ is <1, and the Laplacians of the electron density, $\nabla^2\rho(r)$ (0.04 and 0.2 a.u.), are positive indicating that the interaction is dominated by the contraction of charge away from the interatomic surface toward each nucleus [24–30]. The $\rho(r)$ and $\nabla^2\rho(r)$ at the critical points related to Cr1 ← O10 bond compare well with the respective 0.395 and 1.164 a.u. values reported for the Cr–O bond critical point in the $CrOF_4$ compound [21]. The other important observation, is related to the topological properties of the Cr1 ← O6 and Cr1 ← O10 bond critical points as shown in Table 3.6. In these cases, the electron density values are between 0.4 and 0.5 a.u. while the negative values of the Laplacian of the electron density for the Cr ← O bonds (−0.2 and −0.8 a.u.), observed in Table 3.6, indicate that the Cr1 ← O6 and Cr1 ← O10 bond critical points are not found in a region of charge depletion. The interaction in Cr1 ← O10 bond is the same as the Cr1 ← O6 bond, which has the characteristic of the shared interaction, i.e., the value of electron density at the bond critical point is relatively high and the Laplacian of the charge density is negative indicating that the electronic charge is concentrated in the internuclear region. These values of the Laplacian of the charge density compare well with the -1.096 a.u. values reported for the C–H bond critical point in the VMe_5 compound [21]. Moreover, the (3, +1) critical point as shown in Table 3.6 in chromyl nitrate would confirm the two Cr ← O bonds in the respective structure.

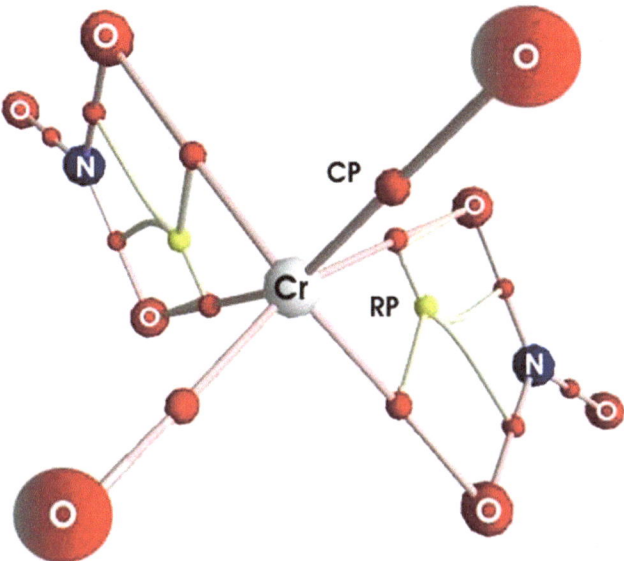

Fig. 3.2 The critical (CP) and ring (RP) points of the charge density for chromyl nitrate. Reprinted from [47], Copyright 2008, with permission from Elsevier

The 12 critical points and the two ring points of the electron density obtained by AIM analysis reveal that the mode of coordination adopted for the nitrate groups in chromyl nitrate is bidentate, as shown in Fig. 3.2. Those above B3LYP level results analyzed for chromyl nitrate are in agreement with the structure observed by electron-diffraction experiments in gas phase and strongly support the conclusions reported previously about the nature of the coordination of the Cr atom for this compound [11].

3.3 Vibrational Study

The structure for the compound has C_2 symmetry and 27 vibrational normal modes. All vibrational modes are IR and Raman active. As it is impossible to make a difference between monodentate and bidentate nitrate groups on the grounds of IR and Raman spectra alone [7, 9], this study was performed taking into account both possibilities. The observed frequencies and the assignment for chromyl nitrate considering the coordination adopted by nitrate groups as monodentate and bidentate are given in Table 3.7. Vibrational assignments were made on the basis of the potential energy distributions (PED) in terms of symmetry coordinates and by comparison with molecules that contain similar groups [1, 31–39].

Table 3.8 shows the calculated harmonic frequencies for chromyl nitrate using B3LYP method with different basis sets. Note that the lowest theoretical frequency

Table 3.7 Experimental frequencies (cm^{-1}) for chromyl nitrate

[a]Experimental spectra				[b]Assignment			[a]Assignment
[a]IR gas	[a]IR liquid	[a]IR solid	[a]Raman liquid	(monodentate)	(bidentate)	(bidentate)[#]	
1900 vvw	1893 vvw	1910		959 + 945 = 1904			1224 + 685 = 1909
1650 sh	1640 sh	1660 sh	1642 (16)	ν_s NO$_2$ ip	ν N=O ip	ν N=O ip	ν_a N=O
1637 vs	1613 vs	1635 vs	1614 sh	ν_s NO$_2$ op	ν N=O op	ν N=O op	ν_s N=O
1556	1550 w	1550					2 × 778 = 1556
		1381					944 + 453 = 1397
		1348					1223 + 125 = 1348
	1305 w	1337		1234 + 100 = 1334			958 + 349 = 1307
	1275	1276		1637 − 349 = 1288			ν_s N=O
1227	1234 sh	1223	1232 (11)	ν_a NO$_2$ ip	ν_a NO$_2$ ip	ν_a NO$_2$ ip	
1221	1215 s		1216 sh	ν_a NO$_2$ op	ν_a NO$_2$ op	ν_a NO$_2$ op	
		1023					775 + 247 = 1022
		968		835 + 125 = 960			
961	958 s	962 sh	959 (100)	ν_s Cr=O	ν_s Cr=O	ν_s Cr=O	ν_a Cr=O
		944	945 sh	ν_a Cr=O	ν_a Cr=O	ν_a Cr=O	ν_s Cr=O
		835		δNO$_2$ ip	δNO$_2$ ip	ν_s NO$_2$ ip	ν N-O
808 sh	806 sh	800		δNO$_2$ op	δNO$_2$ op	ν_s NO$_2$ op	γN=O
781 sh	779 sh	781	782 (8)	γN=O op, γN=O ip	γN=O op, γN=O ip	δNO$_2$ ip, δNO$_2$ op	457 + 349 = 806
776	771 m	774	774 sh	δO=N-O op, δO=N-O ip	ν_s NO$_2$ ip, ν_s NO$_2$ op	γN=O ip, γN=O op	δNO$_2$
685 w	685 w	689	686 (10)	ν_a N-O, ν_s N-O	δO=N-O op, δO=N-O ip	δO=N-O op, δO=N-O ip	δNO$_2$
457 m	457 m	453	460 sh	ν_a Cr-O	δCrO$_2$	δCrO$_2$	ρNO$_2$
			446 (95)	δCrO$_2$	ν_a Cr-O	ν_s Cr ← O	ν_a Cr-O, ν_s Cr-O

(continued)

Table 3.7 (continued)

^aExperimental spectra				^bAssignment			^aAssignment
^aIRgas	^aIRliquid	^aIRsolid	^aRamanliquid	^aRamanliquid	(monodentate)	(bidentate)	(bidentate)#
δCrO_2	349 w		350 (41)		ν_s Cr–O	ν_s Cr–O	ν_s Cr–O
	273 w	271 (9)		Wag CrO_2	τNO_2 op	Wag CrO_2	ρ CrO_2
	247 w	242 (5)		τNO_2 ip	ρ CrO_2	ρ CrO_2	δN–O–Cr
	223 sh			δN–O–Cr op	Wag CrO_2	τw CrO_2	
	213 w	206 (9)		δN–O–Cr ip, ρNO_2 ip	ν_a Cr ← O, δs O–Cr–O	ν_a Cr–O, τN=O	τNO_2
		152 (5)		ρ CrO_2, δO–Cr–O	δa O–Cr–O, τw CrO_2	δa O–Cr–O, δs O–Cr–O	
	125 br, vw	100		τNO_2 op, τw CrO_2, δN–O–Cr op, ρNO_2 op	τN=O, ν_s Cr ← O	τNO_2 ip, ν_a Cr ← O	
	—	—			τNO_2 ip	τNO_2 op	

Abbreviations ν stretching; δ deformation; ρ rocking; wag (γ) wagging; τw torsion; τ torsion; a antisymmetric; s symmetric op, out of phase; ip, in phase

^a Ref. [1]

^b This work

Considering the nitrate group as ring of four members

Reprinted from [47], Copyright 2008, with permission from Elsevier

Table 3.8 Calculated harmonic frequencies (cm^{-1}) for chromyl nitrate using B3LYP method with different basis sets

Exp.	Lanl2DZ	STO-3G	3-21G*	6-31G	6-31+G	6-31+G*	6-31G*	6-311G	6-311G*	6-311+G	6-311+G*
1642	1560	1624	1476	1610	1574	1714	1743	1590	1727	1565	1703
1614	1542	1614	1459	1593	1557	1695	1725	1571	1707	1547	1684
1232	1126	1423.7	1168	1138	1126	1279	1289	1138	1268	1125	1258
1216	1117	1422	1135	1133	1123	1275	1286	1136	1263	1123	1254
959	1076	1248.6	1064	1069	1057	1093	1115	1071	1110	1051	1089
945	1066	1240.5	1048	1069	1054	1085	1111	1069	1104	1045	1079
835	869	964.7	846	859	859	987	991	867	975	863	972
800	861	957.8	842	854	853	982	986	863	970	859	966
781	712	722.5	723	732	733	784	785	732	794	728	793
781	710	699.9	721	730	731	784	783	729	792	727	793
774	682	583.5	662	710	719	777	778	705	785	714	785
774	681	578.5	662	710	718	776	777	705	784	714	783
686	618	560.9	607	626	626	683	684	629	695	624	691
686	617	540.6	601	622	623	682	684	626	693	621	690
460	450	539.7	476	456	448	467	476	463	482	447	467
446	431	434.8	456	443	438	451	453	444	459	437	450
349	345	387.2	365	346	341	351	356	348	359	339	350
273	281	335.3	286	280	277	284	287	286	293	273	283
242	255	311.7	279	265	258	269	274	264	274	257	266
223	225	305	244	231	225	240	247	237	248	226	237
206	191	300.9	242	211	198	220	236	204	220	201	211
206	183	240.4	214	197	192	200	207	197	207	190	197
152	173	228.5	177	176	172	186	191	179	188	173	181
152	144	160.9	151	144	143	148	149	146	152	141	147
100	97	116.7	99	99	97	105	111	101	106	96	103
100	87	105.9	93	87	79	101	105	86	100	81	93
	70	80.1	73	71	67	73	76	71	75	67	72
RMSD	59.89	126.8	69.9	49.8	50.9	58.8	63.7	51.6	61.9	46.0	53.6

was not observed in the vibrational spectrum and for this reason these frequencies were taken as experimental values. In all cases, the theoretical values were compared with the respective experimental values by means of the RMSD values. It can be seen that the best results are obtained with a B3LYP/6-311+G calculation and that the introduction of diffuse functions (but not of polarization functions!) is essential to have a good approximation to the experimental values, especially in the case of the Cr=O and Cr–O stretchings. The results obtained at B3LYP level with 6-31G* basis set were considered because after scaling this method presents satisfactory agreement between the calculated and the experimental vibrational frequencies of chromyl nitrate. In general, the theoretical IR and Raman spectra of the chromyl nitrate demonstrate good agreement with the experimental spectrum, especially in the higher intensity of the Cr=O stretching bands (see Figs. 3.3 and 3.4). It is possible to observe that in all calculations some vibrational modes of different symmetries are mixed among them because the frequencies are approximately the same.

Bellow, a discussion of the assignment of the most important groups for the compound studied considering the two coordination kinds is presented.

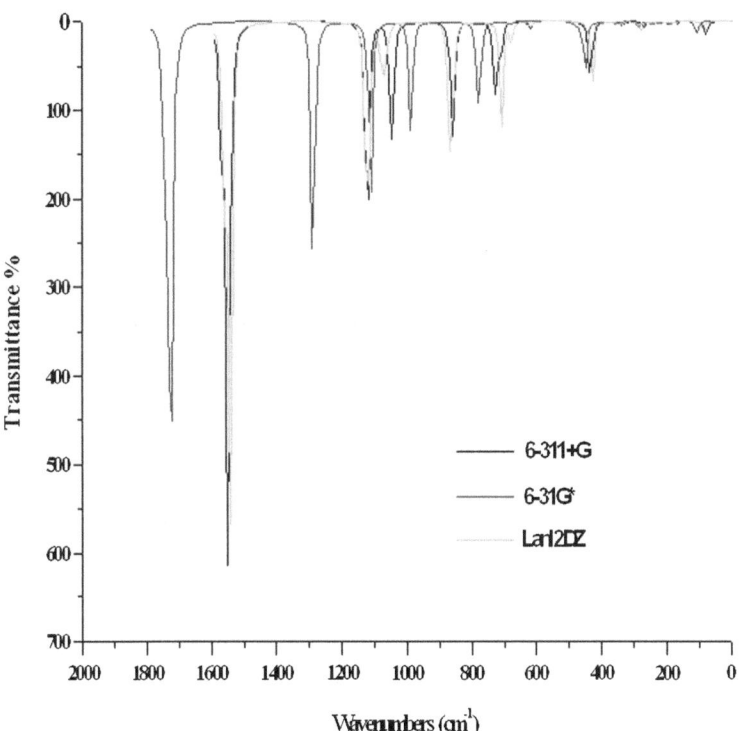

Fig. 3.3 Theoretical IR spectrum of $CrO_2(NO_3)_2$ at different levels of theory. Reprinted from [47], Copyright 2008, with permission from Elsevier

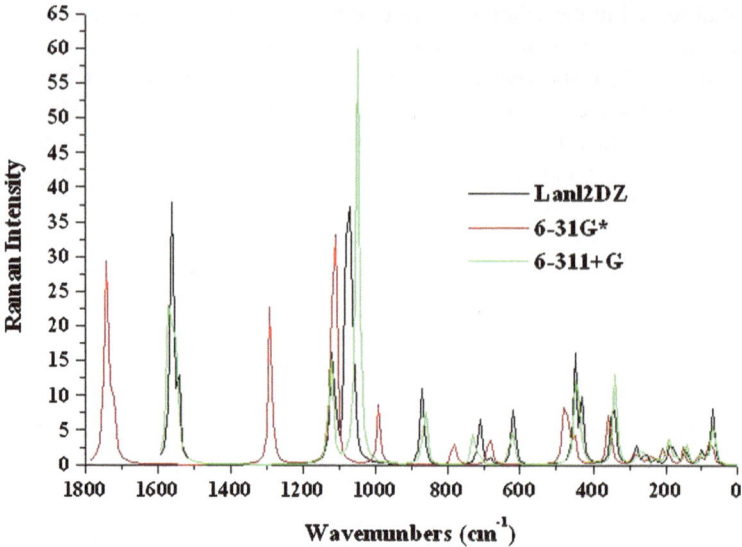

Fig. 3.4 Theoretical Raman spectrum of $CrO_2(NO_3)_2$ at different levels of theory. Reprinted from [47], Copyright 2008, with permission from Elsevier

3.4 Monodentate Coordination of the Nitrate Groups

The IR and Raman frequencies, their respective intensities and the potential energy distribution obtained by DFT/B3LYP/6-31G*, B3LYP/Lanl2DZ, and B3LYP/6-311+G calculations considering monodentate coordination of the nitrate groups appear in Tables 3.7, 3.9, and 3.10, respectively. In all cases, the theoretical values were compared with the experimental values by means of the RMSD values. The frequencies calculated using the 6-311+G basis set for the compound are lower than those obtained using the Lanl2DZ and 6-31G* basis sets.

The RMSD initial values using B3LYP/Lanl2DZ and B3LYP/6-31G* calculations are 62.2 and 73.5 cm^{-1} while value with the 6-311+G basis set is 53.1 cm^{-1}. It can be seen that the best results for chromyl nitrate are obtained using B3LYP/6-31G* calculation with a RMSD final equal at 19.8 cm^{-1} although the introduction of the polarization function does not have a good approximation to the experimental values. In this case, the covalent bonding of the nitrate group is easily recognized from its IR spectrum because the symmetry group changes from the point group D_{3h} of the free ion to C_{2v} of the compound [7].

Table 3.9 Observed and calculated wavenumbers (cm^{-1}), potential energy distribution, and assignment for monodentate chromyl nitrate

Mode	Observed[a]	Calculated[b]	SQM[c]	IR int.[d]	Raman act.[e]	PED ($\geq$10 %)
Modes A symmetry						
1	1642	1560	1628	180.0	37.0	64 S_1 + 16 S_2
2	1216	1117	1233	138.0	21.0	58 S_2 + 27 S_1
3	959	1076	954	60.4	43.6	94 S_3
4	835	869	833	129.3	10.9	47 S_4 + 34 S_{21}
5	781	712	783	32.5	5.1	74 S_5 + 10 S_{12}
6	774	682	770	0.9	0.4	33 S_9 + 33 S_6 + 14 S_{11} + 13 S_5
7	686	618	624	1.2	7.5	42 S_7 + 22 S_{18}
8	460	450	483	3.6	15.5	53 S_8 + 17 S_4 + 15 S_{20}
9	349	345	375	3.7	14.4	47 S_{22} + 29 S_9 + 11 S_{12}
10	242	255	250	0.0	2.5	30 S_{10} + 27 S_{13} + 23 S_{12} + 14 S_{14}
11	206	191	190	1.2	1.8	38 S_{11} + 17 S_{13} + 13 S_9 + 10 S_{12}
12	206	183	183	0.0	2.5	29 S_{11} + 27 S_{12} + 15 S_{10} + 10 S_{14}
13	152	144	142	0.1	3.1	27 S_{12} + 25 S_{14} + 21 S_{13} + 20 S_{10}
14	70	70	70	0.0	8.0	50 S_{10} + 43 S_{12}
Modes B symmetry						
15	1614	1542	1618	687.0	13.3	66 S_{15} + 16 S_{16}
16	1232	1126	1247	215.0	0.9	59 S_{16} + 26 S_{15}
17	945	1066	950	56.1	32.3	94 S_{17}
18	800	861	831	77.4	0.5	46 S_{18} + 34 S_7 + 16 S_6
19	781	710	782	91.9	2.3	43 S_{19} + 21 S_{20} + 20 S_8
20	774	681	768	20.8	0.6	31 S_{19} + 27 S_{20} + 22 S_8
21	686	617	615	4.3	1.9	51 S_{21} + 15 S_4 + 11 S_8
22	446	431	454	68.7	9.1	61 S_{22} + 12 S_{12}
23	273	281	271	11.3	2.5	52 S_{23} + 25 S_{27} + 15 S_{26}
24	223	225	219	4.0	0.7	29 S_{23} + 25 S_{27} + 17 S_{24} + 16 S_{26}
25	152	173	171	1.0	0.9	23 S_{24} + 21 S_{27} + 21 S_{25} + 14 S_{23}
26	100	97	96	0.3	2.5	68 S_{26} + 23 S_{27}
27	100	87	85	9.9	0.1	28 S_{24} + 28 S_{25} + 13 S_8
RMSD		62.2	22.4			

[a] This work
[b] DFT B3LYP/Lanl2DZ
[c] From scaled quantum mechanics force field
[d] Units are km mol^{-1}
[e] Raman activities in Å^4 (amu)$^{-1}$ RMSD (cm^{-1})

3.5 Nitrate Groups

The strong bands observed in the IR spectrum reported in previous paper [1] of CrO$_2$(NO$_3$)$_2$ liquid at 1,613 and 1,215 cm^{-1}, were assigned to N=O antisymmetric and symmetric stretchings, respectively. In this work, the N=O in-phase and out-of-phase symmetric stretching modes are calculated by the B3LYP/6-31G* method at 1,743 and 1,724 cm^{-1}, respectively. The B3LYP/Lanl2DZ and B3LYP/

Table 3.10 Observed and calculated wavenumbers (cm^{-1}), potential energy distribution, and assignment for monodentate chromyl nitrate

Mode	Observed[a]	Calculated[b]	SQM[c]	IR int.[d]	Raman act.[e]	PED ($\geq$10 %)
Modes A symmetry						
1	1642	1565	1626	207.3	40.0	60 S_1 + 20 S_2 + 10 S_6
2	1216	1123	1233	129.5	19.6	55 S_2 + 30 S_1
3	959	1051	953	78.4	45.2	94 S_3
4	835	863	839	107.8	9.9	42 S_4 + 34 S_{21} + 11 S_8
5	781	728	783	35.6	2.7	64 S_5 + 10 S_9 + 10 S_6
6	774	714	770	0.0	0.3	36 S_6 + 22 S_9 + 21 S_5 + 10 S_{11}
7	686	624	624	3.3	6.0	45 S_7 + 23 S_{18} + 12 S_9
8	460	447	487	4.1	16.3	56 S_8 + 15 S_4 + 15 S_{20}
9	349	339	366	4.1	13.6	49 S_{22} + 28 S_9 + 13 S_{12}
10	242	257	252	0.0	1.8	35 S_{10} + 24 S_{13} + 22 S_{12} + 11 S_{14}
11	206	201	199	1.9	1.5	42 S_{12} + 20 S_{11} + 11 S_{10} + 10 S_{13}
12	206	190	191	0.6	3.2	41 S_{11} + 17 S_{13} + 13 S_9 + 10 S_{12}
13	152	141	139	0.0	3.1	29 S_{10} + 26 S_{12} + 21 S_{13} + 20 S_{14}
14	67	67	68	0.1	7.5	49 S_{10} + 41 S_{12}
Modes B symmetry						
15	1614	1547	1616	744.1	17.6	62 S_2 + 19 S_{15}
16	1232	1124	1248	193.3	0.7	57 S_{16} + 28 S_{15}
17	945	1045	950	102.1	29.5	94 S_{17}
18	800	858	832	53.4	0.4	47 S_{18} + 37 S_7 + 11 S_6
19	781	727	783	60.0	2.4	66 S_{19} + 13 S_{20}
20	774	714	767	42.7	1.0	38 S_{20} + 21 S_8 + 16 S_4 + 13 S_{19}
21	686	621	616	6.8	1.1	53 S_{21} + 16 S_4 + 14 S_8
22	446	437	448	77.7	6.5	56 S_{22} + 14 S_{12} + 12 S_9
23	273	273	268	7.7	2.3	50 S_{23} + 24 S_{27} + 17 S_{26}
24	223	226	220	3.5	0.7	31 S_{23} + 21 S_{24} + 19 S_{27} + 15 S_{26}
25	152	173	170	2.1	1.1	23 S_{27} + 21 S_{25} + 20 S_{24} + 15 S_{23} + 10 S_{26}
26	100	96	97	0.9	2.4	52 S_{26} + 24 S_{27} + 17 S_{25}
27	100	81	78	14.9	0.2	30 S_{24} + 27 S_{25} + 12 S_8
RMSD		53.0	22.0			

[a] This work
[b] DFT B3LYP/6-311+G
[c] From scaled quantum mechanics force field
[d] Units are km mol^{-1}
[e] Raman activities in Å^4 (amu)$^{-1}$ RMSD (cm^{-1})

6-311+G methods underestimate the N=O stretching frequencies as compared to the experimental values. The frequencies predicted for these vibrational modes show that the two symmetric stretching modes are split by about 19 cm^{-1} with the 6-31G* basis set (using Lanl2DZ and 6-311+G basis sets, the splitting is 18 cm^{-1}). These modes are calculated slightly coupled with antisymmetric stretching modes but they are only mixed with the O=N–O in-phase deformation mode using 6-31G* and 6-311+G basis sets. Normally, the N=O stretching frequencies of the nitrate ion and nitrate group are observed between 1,531 and 1,481 cm^{-1} [31, 36, 37] while in nitrogen oxides they are observed at 1,758 and 1,660 cm^{-1} [8]. In $[Zr(NO_3)_6]^{2-}$ complex [32], the N=O stretching is observed at 1,570 cm^{-1}, in $[Cr(NH_3)_5(ONO)]Cl_2$ complex at 1,460 cm^{-1} [36] while this vibrational mode in HNO_3 [32] is observed at 1,672 cm^{-1}. In the $UO_2(NO_3)_2$ and $Cu(NO_3)_2$ anhydrous salts, the N=O stretching frequencies are observed at 1,560 and 1,585 cm^{-1} [34, 35]. For these observations, the shoulder at 1,640 cm^{-1} in the liquid spectrum and the very strong band at 1,613 cm^{-1}, with a difference between them of 27 cm^{-1}, are assigned to these modes.

The NO_2 in-phase and out-of-phase antisymmetric modes are calculated with B3LYP/6-31G* method at 1,289 and 1,285 cm^{-1}, respectively. In the three calculations, these modes appear coupled with the stretching modes. Generally, in nitro complexes the NO_2 antisymmetric stretching modes are observed between 1,488 and 1,343 cm^{-1} and in nitrate ion is observed at 1,388 cm^{-1} [8, 36] while the symmetric mode is observed between 1,364 and 1,315 cm^{-1} [36]. In the $UO_2(NO_3)_2$ and $Cu(NO_3)_2$ anhydrous salts, the NO_2 stretching antisymmetric modes are observed in the 1,300 cm^{-1} region while the NO_2 stretching symmetric modes appear approximately at 1,000 cm^{-1} [34, 35]. In our previous paper [1], only the shoulder in the Raman spectrum at 945 cm^{-1} is assigned to one of these modes. Now, the shoulder in the IR spectrum of the liquid phase at 1,234 cm^{-1} and the strong band in the same spectrum at 1,215 cm^{-1} are assigned to NO_2 in-phase and out-of-phase antisymmetric stretching modes, respectively.

The NO_2 out-of-phase and in-phase deformation modes are perfectly characterized by the three calculations and are observed coupled with N–O vibrational modes. The bands observed in the IR spectrum at low temperature at 835 and 800 cm^{-1} are assigned to NO_2 in-phase and out-of-phase deformation, respectively. These bands are predicted in the Raman spectrum with all basis sets with low intensity and experimentally are not observed.

Other modes well characterized by the three calculations are the N=O out-of-plane in-phase and out-of-phase deformation. The PED values indicate that the O=N–O in-phase and out-of-phase deformation modes are strongly coupled with vibrations of the CrO_2 and NO_2 groups. In nitro complexes, these modes are observed between 657 and 433 cm^{-1} [36] while in nitrate ion is observed at 831 cm^{-1} [8, 36]. In this case, the band in the low temperature IR at 781 cm^{-1} is assigned to these vibrational modes. The band in the low temperature IR observed at 774 cm^{-1} in the previous paper is assigned to these vibrational modes [1]. In this case, the calculations confirm such assignment.

The N–O antisymmetric and symmetric stretching modes calculated are also coupled with vibrations of nitrate groups as shown in Tables 3.7, 3.8, and 3.9. In nitrate ion, these modes are observed at 831 cm^{-1} [8, 36] while in $UO_2(NO_3)_2$ they appear at 800 cm^{-1} [34]. The weak band in the Raman spectrum at 686 cm^{-1}, previously assigned to nitrate rocking mode [1], is assigned in this case to these vibrational modes.

In the region of lower frequencies, the vibrational modes normally expected for the nitrate groups are the N–O–Cr bending, rocking, wagging, and twisting modes as observed in covalent nitrates [32]. All modes, as shown in Tables 3.7, 3.9, and 3.10, appear coupled among them and with other modes of the chromyl group while the PED values are different in the calculations with all the basis sets.

Experimentally, in the nitro complexes the NO_2 rocking and twisting modes are observed in the low frequency region between 600 and 400 cm^{-1} and 300 and 240 cm^{-1}, respectively [33, 36, 37]. In a previous paper, we assigned the NO_2 rocking mode at 686 cm^{-1}, the NO_2 torsion at 152 cm^{-1} while the CrO_2 twisting mode was not assigned [1]. In chromyl nitrate, the NO_2 in-phase and out-of-phase torsion modes are calculated using the three basis sets which appear strongly mixed and with different PED. Previously, we assigned only one of these modes at 152 cm^{-1} [1]. In this work, with the aid of the calculations the very weak band in the Raman spectrum at 242 cm^{-1} is assigned to NO_2 in-phase torsion mode while the band in the same spectrum at 100 cm^{-1} is assigned to NO_2 out-of-phase torsion mode.

The assignment of the N–O–Cr in-phase and out-of-phase bending modes is very difficult because these modes are calculated by three methods used at different frequencies and PED values. With the B3LYP/6-31G* calculation, these modes appear more defined than with the other methods hence the shoulder observed in the IR spectrum of the liquid compound at 223 cm^{-1} and the weak Raman band at 206 cm^{-1} are assigned to N–O–Cr out-of-phase and in-phase modes bending modes, respectively. In our assignment previously realized for this molecule, only one of these modes was assigned at 247 cm^{-1} [1].

The NO_2 in-phase rocking mode is calculated clearly at 207 cm^{-1} with the B3LYP/6-31G* calculation and with higher PED value (55 %) than with the B3LYP/6-311+G calculation (42 %) but, in this last case, it is calculated at 201 cm^{-1}. The NO_2 out-of-phase rocking mode appears mixed, in the three methods used, with different modes such as NO_2 torsion and CrO_2 twisting. This mode, with the B3LYP/6-31G* method, is calculated clearly at 112 cm^{-1} coupled with higher PED value (27 %) with the CrO_2 twisting mode while with the B3LYP/6-311+G method it is calculated at 81 cm^{-1} with higher contribution (27 %) and, it is also calculated at 173 cm^{-1} but, with lower contribution (21 %). Experimentally, in nitro complexes the NO_2 rocking mode is observed in the low frequency region, between 600 and 400 cm^{-1} [36]. In a previous paper, we assigned the NO_2 rocking mode at 686 cm^{-1} [1]. In this case, the theoretical calculations show clearly the NO_2 in-phase rocking mode and for this reason the band at 206 cm^{-1} is also assigned to this vibrational mode and to the N–O–Cr in-phase bending mode. The band in the Raman spectrum at 100 cm^{-1} is assigned to the NO_2 out-of-phase rocking mode.

3.6 Chromyl Group

The frequencies predicted for the vibrational modes of chromyl nitrate show that the antisymmetric and symmetric Cr=O stretchings are split by more than 4 cm^{-1}, indicating a little contribution of the central Cr atom in these vibrations. The antisymmetric and symmetric Cr=O stretchings modes were observed in the spectrum of the solid sample at 968 and 962 cm^{-1}, respectively, while the more intense band at 959 cm^{-1} in the Raman spectrum is assigned to the symmetric Cr=O stretching. Tables 3.7, 3.9, and 3.10 for chromyl nitrate show that the unscaled DFT frequencies for the symmetric Cr=O stretchings mode, are higher than the frequencies of the antisymmetric Cr=O stretchings, an observation also reported by us [1]. In this compound, these modes are uncoupled with other modes. In other chromyl compounds, these modes appear in 1050–900 cm^{-1} region, i.e., in $CrO_2(ClO_4)_2$ they appear at 990 and 980 cm^{-1} [38], in $CrO_2(SO_3F)_2$ appear at 1,061 and 1,020 cm^{-1} [14] and in CrO_2F_2 and CrO_2Cl_2 they are observed for the first compound at 1,016 and 1,006 cm^{-1} while for the second one at 1,002 and 995 cm^{-1}, respectively [32, 36]. In this case, the intense band in the Raman spectrum at 959 cm^{-1} is assigned to Cr=O symmetric stretching mode while the shoulder observed in the same spectrum at 945 cm^{-1} is assigned to the corresponding antisymmetric stretching.

Also, the antisymmetric and symmetric Cr–O stretchings are split by more than 26 cm^{-1}, indicating a slight contribution of the central Cr atom in these vibrations. These stretchings in $CrO_2(ClO_4)_2$ are observed, respectively, at 380 and 355 cm^{-1} [38] and in $CrO_2(NO_3)_2$ were assigned previously by us at 460 and 446 cm^{-1}, respectively [1]. In this case, the theoretical calculation predicts these modes with greater PED value for the antisymmetric mode in reference to the symmetric mode. The intensities of these bands using all basis sets are not predicted correctly because the more intense band is related to the antisymmetric mode. Here, the previous assignment for the antisymmetric mode at 460 cm^{-1} [1] is confirmed while the band at 350 cm^{-1} of the higher intensity in the Raman spectrum is assigned to Cr–O symmetric stretching mode. The CrO_2 bending mode is observed in CrO_2F_2 at 364 cm^{-1} while in CrO_2Cl_2 at 356 cm^{-1} [32, 36]. The band observed in the Raman spectrum at 350 cm^{-1} is assigned to the CrO_2 bending (O=Cr=O) of chromyl nitrate [1] while the other CrO_2 bending (O–Cr–O) was not assigned. In this work, the B3LYP/6-31G* method calculates the CrO_2 bending mode at 453 cm^{-1} with 63 % of contribution PED and at 437 cm^{-1} with 56 % of contribution PED. With the other basis set, this mode appears to be also coupled. In this case, the intense band in the Raman spectrum at 446 cm^{-1} is assigned to CrO_2 bending. The other O–Cr–O bending mode is calculated to be (at 149 cm^{-1} with 6-31G* basis set and at 141 cm^{-1} with 6-311+G basis set) strongly coupled with other modes. The very weak band observed in the Raman spectrum at 152 cm^{-1} is assigned to O–Cr–O bending. Previously, this mode was not assigned [1].

The wagging, rocking, and twisting modes of the CrO_2 group are not assigned in a previous paper [1]. In this case, the calculations predict these modes in the low frequency region and they are coupled with other modes of the nitrate groups. The wagging CrO_2 mode is calculated using all basis sets at higher frequency and with lower contribution of the PED than the rocking mode. For these observations, the weak band in the spectrum of the liquid at 273 cm^{-1} is assigned to the wagging mode while the Raman band at 152 cm^{-1} is assigned to the rocking mode.

The CrO_2 twisting mode was not assigned previously. The PED values indicate that this mode is strongly coupled with vibrations of the same group and the NO_2 group as shown in Tables 3.7, 3.9, and 3.10. It is noticeable how the contribution of the PED value changes with the method used and it is possible to observe with all basis sets that this mode is strongly mixed at different frequencies (between 273 and 100 cm^{-1}). In this case, this mode could be assigned at 100 cm^{-1} because it appears with a higher PED value.

3.7 Bidentate Coordination of the Nitrate Groups

3.7.1 Nitrate Group Without Ring

The frequencies, IR and Raman intensities, and potential energy distribution obtained by B3LYP/6-31G*, B3LYP/Lanl2DZ, and B3LYP/6-311+G calculations considering the mode of coordination adopted by nitrate groups as bidentate appear in Tables 3.7, 3.11, and 3.12. In the three cases, the comparison of the theoretical values with the respective experimental values (RMSD) is observed in the respective Tables. It can be seen that the best results for bidentate coordination of chromyl nitrate are newly obtained with B3LYP/6-31G* calculation with a final RMSD of 10.6 cm^{-1} while with the Lanl2DZ and 6-311+G basis sets the final RMSDs were 18.7 and 16.5 cm^{-1}, respectively.

3.7.2 Nitrate Group

The theoretical results show slight changes in the PED values and in the coupling of the modes. In this case, using the two basis sets the N=O stretchings in-phase mode, they appear coupled with the NO_2 deformation in-phase mode and, on the other hand, the corresponding out-of-phase modes are also coupled among them. This way, the shoulder in the spectrum of liquid phase at 1,640 cm^{-1} and the very strong band in the same spectrum at 1,613 cm^{-1} are assigned to these modes. Both bands in the Raman spectrum are observed at 1,642 and 1,614 cm^{-1}.

The NO_2 antisymmetric out-of-phase and in-phase frequencies also appear coupled with the corresponding O=N–O out-of-phase and in-phase deformation modes, respectively but, with greater contribution (in the two modes) when the 6-31G* basis set is used. The assignment of these bands is similar to the mono-dentate type, where the shoulder at 1,234 cm^{-1} and the strong band at 1,215 cm^{-1} in the IR spectrum of the liquid phase are assigned to NO_2 antisymmetric in-phase and out-of-phase modes, respectively.

The NO_2 in-phase and out-of-phase deformation modes are also coupled with vibrations of the nitrate groups as shown in Tables 3.7, 3.11, and 3.12. In this case,

Table 3.11 Observed and calculated wavenumbers (cm^{-1}), potential energy distribution, and assignment for bidentate chromyl nitrate

Mode	Observed[a]	Calculated[b]	SQM[c]	IR int.[d]	Raman act.[e]	PED ($\geq$10 %)
Modes A symmetry						
1	1642	1560	1663	180.0	37.0	72 S_1 + 16 S_4
2	1216	1117	1176	138.0	21.0	64 S_{16} + 29 S_{21}
3	959	1076	952	60.4	43.6	93 S_3
4	835	869	840	129.3	10.9	43 S_4 + 19 S_{20} + 11 S_{21}
5	781	712	783	32.5	5.1	71 S_5 + 27 S_{27}
6	774	682	765	0.9	0.4	75 S_{19} + 10 S_{20}
7	686	618	658	1.2	7.5	30 S_2 + 28 S_{20} + 22 S_7
8	460	450	467	3.6	15.5	67 S_8 + 20 S_9
9	349	345	363	3.7	14.4	34 S_9 + 33 S_8 + 11 S_{11}
10	242	255	247	0.0	2.5	28 S_{25} + 24 S_{10} + 18 S_{11}
11	206	191	213	1.2	1.8	49 S_{11} + 34 S_9
12	206	183	188	0.0	2.5	68 S_{26} + 23 S_{25}
13	152	144	137	0.1	3.1	54 S_{10} + 31 S_{26}
14	70	70	70	0.0	8.0	62 S_{14} + 32 S_{25}
Modes B symmetry						
15	1614	1542	1645	687.0	13.3	73 S_{15} + 16 S_{18}
16	1232	1126	1186	215.0	0.9	65 S_2 + 29 S_7
17	945	1066	952	56.1	32.3	93 S_{17}
18	800	861	827	77.4	0.5	48 S_{18} + 23 S_6
19	781	710	773	91.9	2.3	25 S_6 + 19 S_{18} + 18 S_{21} + 14 S_{22} + 10 S_7
20	774	681	760	20.8	0.6	26 S_{20} + 25 S_{19} + 19 S_4 + 11 S_9 + 10 S_7
21	686	617	653	4.3	1.9	33 S_{16} + 30 S_6 + 16 S_{21}
22	446	431	447	68.7	9.1	29 S_{22} + 28 S_{24} + 12 S_{21} + 10 S_{27}
23	273	281	285	11.3	2.5	38 S_{24} + 26 S_{12} + 22 S_{23}
24	223	225	219	4.0	0.7	25 S_{22} + 18 S_{13} + 17 S_{23} + 16 S_{24} + 16 S_{12}
25	152	173	142	1.0	0.9	27 S_{24} + 26 S_{12} + 15 S_{13} + 12 S_{27}
26	100	97	103	0.3	2.5	56 S_{27} + 28 S_{24}
27	100	87	89	9.9	0.1	52 S_{23} + 34 S_{27}
RMSD		62.2	18.7			

[a] This work
[b] DFT B3LYP/Lanl2DZ
[c] From scaled quantum mechanics force field
[d] Units are km mol^{-1}
[e] Raman activities in Å^4 (amu)$^{-1}$ RMSD (cm^{-1})

the NO$_2$ in-phase mode is calculated with 6-31G* basis set at 991 cm^{-1} and the corresponding out-of-phase mode at 986 cm^{-1} while with 6-311+G basis set they are calculated to be strongly coupled and with a higher contribution of PED at 863 and 858 cm^{-1}, respectively. The bands in the low temperature IR spectrum observed at 835 and 800 cm^{-1} are assigned to these vibrational modes.

Table 3.12 Observed and calculated wavenumbers (cm^{-1}), potential energy distribution, and assignment for bidentate chromyl nitrate

Mode	Observed[a]	Calculated[b]	SQM[c]	IR int.[d]	Raman act.[e]	PED ($\geq$10 %)
Modes A symmetry						
1	1642	1565	1662	207.3	40.0	72 S_1 + 17 S_4
2	1216	1123	1179	129.5	19.6	63 S_2 + 31 S_{21}
3	959	1051	953	78.4	45.2	95 S_3
4	835	863	832	107.8	9.9	51 S_4 + 18 S_{20} + 10 S_{21}
5	781	728	783	35.6	2.7	30 S_6 + 15 S_{18} + 15 S_{21} + 13 S_5 + 10 S_{22}
6	774	714	774	0.0	0.3	35 S_7 + 19 S_{19} + 17 S_7 + 14 S_4
7	686	624	662	3.3	6.0	34 S_2 + 31 S_{20} + 23 S_7
8	460	447	464	4.1	16.3	71 S_8 + 19 S_9
9	349	339	354	4.1	13.6	38 S_9 + 29 S_8 + 12 S_{11}
10	242	257	251	0.0	1.8	29 S_{25} + 28 S_{10} + 17 S_{14} + 16 S_{11}
11	206	201	212	1.9	1.5	45 S_{11} + 32 S_9
12	152	173	149	0.6	3.2	33 S_{24} + 32 S_{27} + 13 S_{23} + 11 S_{12}
13	100	96	99	0.9	2.4	31 S_{13} + 20 S_{12} + 19 S_{22}
14	67	67	68	0.1	7.5	60 S_{14} + 36 S_{25}
Modes B symmetry						
15	1614	1547	1644	744.1	17.6	72 S_{15} + 17 S_{18}
16	1232	1124	1188	193.3	0.7	64 S_{16} + 30 S_7
17	945	1045	952	102.1	29.5	94 S_{17}
18	800	858	824	53.4	0.4	54 S_{18} + 21 S_6
19	781	727	777	60.0	2.4	68 S_5 + 25 S_{27}
20	774	714	769	42.7	1.0	78 S_{19}
21	686	621	656	6.8	1.1	36 S_{16} + 32 S_6 + 19 S_{21}
22	446	437	448	77.7	6.5	34 S_{24} + 25 S_{22} + 11 S_{27}
23	273	273	280	7.7	2.3	39 S_{24} + 24 S_{23} + 23 S_{12}
24	223	226	220	3.5	0.7	24 S_{24} + 20 S_{12} + 17 S_{22} + 15 S_{23} + 14 S_{13}
25	206	190	196	2.1	1.1	37 S_{26} + 21 S_{11} + 18 S_{25} + 12 S_9
26	152	141	137	0.0	3.1	53 S_{10} + 28 S_{26}
27	100	81	95	14.9	0.2	45 S_{27} + 33 S_{23}
RMSD		53.0	16.5			

[a] This work
[b] DFT B3LYP/6-311+G
[c] From scaled quantum mechanics force field
[d] Units are km mol^{-1}
[e] Raman activities in Å^4 (amu)$^{-1}$ RMSD (cm^{-1})

The two N=O out-of-plane deformation modes (out-of-phase and in-phase) appear strongly coupled when the 6-311+G basis set is used (at 728 and 727 cm^{-1}, respectively). The N=O in-phase deformation mode appears uncoupled (6-31G* basis set) at 783 cm^{-1} and with a percentage of 83 % in the PED values while the corresponding out-of-phase deformation mode is calculated at 785 cm^{-1} (with 61 % PED) slightly coupled with the N=O torsion mode.

In both cases, the scaled frequencies are 781 and 780 cm^{-1}, respectively, with 6-31G* basis set and 783 and 777, respectively, with 6-311+G basis. Hence, the band in the low temperature IR at 781 and at 782 cm^{-1} in the Raman spectrum is assigned to these vibrational modes like in the monodentate type. The two calculations confirm such an assignment.

An important observation in this case is the great difference in frequency that appears between the corresponding NO_2 symmetric out-of-phase and in-phase modes. Particularly, the out-of-phase mode is strongly coupled with the NO_2 out-of-phase deformation mode and has lower contribution but, only for the B3LYP/6-31G* calculation. For this last basis set, the calculated frequencies are 785 and 777 cm^{-1}, respectively. On the contrary, the NO_2 symmetric out-of-phase and in-phase modes are calculated with 6-311+G basis set with higher percentage PED than the other basis set and both modes are calculated at the same frequency (714 cm^{-1}). The shoulder in the Raman spectrum of the liquid phase at 774 cm^{-1} is assigned to these vibrational modes.

The O=N–O out-of-phase and in-phase deformation modes are also coupled and they are calculated at 618 and 617 cm^{-1} with Lanl2DZ basis set while they appear at 624 and 621 cm^{-1}, with 6-31G* and 6-311+G basis sets, respectively. For this observation, the weak band in the Raman spectrum at 686 cm^{-1} is assigned to these vibrational modes. In the monodentate type, these modes are assigned at 774 cm^{-1}.

In the low frequency region, the vibrational modes of the nitrate group appear strongly coupled with other modes of the chromyl group as shown in Tables 3.7, 3.11, and 3.12. When the nitrate groups present bidentate coordination other vibrational modes, besides the N=O torsion mode, as NO_2 out-of-phase and in-phase torsion modes are observed in this region.

The two calculations characterize perfectly the NO_2 in-phase torsion mode with a 54 % of contribution PED (6-31G* basis set). The corresponding out-of-phase mode is observed mixed with different vibrational modes but, the higher contribution to PED is at 287 cm^{-1} (6-31G* basis set). Both torsion modes are observed coupled with the vibrational modes of the chromyl group at 76 and 273 cm^{-1}, respectively. Similarly for the monodentate type, the very weak band in the Raman spectrum at 271 cm^{-1} is assigned to the NO_2 torsion out-of-phase mode while the NO_2 torsion in-phase mode could not be assigned because the lower frequency is not observed in the vibrational spectra for the compound or probably is overlapped with the other bands.

The N=O torsion mode is calculated with higher contribution at 105 cm^{-1} (6-31G* basis set) and 81 cm^{-1} when the size basis set increases. Thus, the observation of a weak band in the Raman spectrum of the liquid compound at 100 cm^{-1} could also be assigned to this mode as observed in Table 3.7.

3.7.3 Chromyl Group

For chromyl nitrate, it is possible to observe two Cr=O stretching modes and four Cr–O stretching modes due to bidentate coordination of the nitrate groups. The scaled DFT frequencies for the Cr=O antisymmetric and symmetric stretching frequencies are in good agreement with the experimental frequencies and normal coordinate calculations. The Cr=O antisymmetric and symmetric stretching modes are easily assigned by comparison with the calculations because in the three cases studied they appear with a higher contribution to PED and without coupling as shown in Tables 3.7, 3.11, and 3.12. Moreover, in chromyl compounds these modes are observed in 1050–900 cm^{-1} region [31, 32, 36, 38]. The assignment of these modes is similar to the monodentate type as shown in Table 3.7.

One important observation is that the CrO_2 bending (O=Cr=O) mode appears at higher frequencies than the monodentate type due to four Cr–O bonds: two Cr–O bonds and two Cr $\leftarrow$ O bonds (Fig. 3.2). Hence, it is possible to observe this mode at 475 cm^{-1} with a contribution to PED of 72 % using 6-31G* basis set. In all calculations, this mode appears slightly coupled with the Cr–O symmetric stretching. Hence, the shoulder observed in the Raman spectrum at 460 cm^{-1} is assigned to CrO_2 bending (O=Cr=O).

The theoretical calculations predict approximately the Cr–O symmetric stretching mode with the same percentage PED value (38 %) coupled with the CrO_2 deformation and Cr–O antisymmetric stretching modes. The antisymmetric mode is calculated at lower frequencies and both modes with approximately the same contribution (44 %). The symmetric mode calculated with higher intensity Raman was assigned to the intense band observed in the Raman spectrum at 446 cm^{-1} while the corresponding antisymmetric stretching was assigned to the weak band observed in the IR spectrum of the liquid compound at 349 cm^{-1}.

The antisymmetric and symmetric Cr $\leftarrow$ O stretchings are obviously calculated at lower frequencies because bond lengths are greater than the Cr–O bonds (see Table 3.7). It is possible to observe the symmetric mode with higher contribution to PED at 112 cm^{-1} (26 %) with 6-31G* basis set. The antisymmetric stretchings are observed with lower contribution (25 %) mixed with other vibrational modes at higher frequencies (453 cm^{-1}) with 6-31G* basis set. These modes were assigned in accordance to a higher contribution; i.e., at 206 and 100 cm^{-1}, the higher frequency correspond to the antisymmetric stretchings.

In all calculations, it is possible to observe the wagging, rocking, and twisting modes of the CrO_2 group strongly mixed with other modes. Differently from the monodentate type, the CrO_2 wagging mode is calculated, using 6-31G* basis set, at a lower frequency but, with higher contribution to PED (247 cm^{-1}, 42 % PED), while the opposite occurs for the CrO_2 rocking mode (calculated at 274 cm^{-1} (34 %) with 6-31G* basis set). In this case, the shoulders observed at 247 and 223 cm^{-1} in the IR spectrum of the liquid chromyl nitrate are assigned to CrO_2 rocking and wagging modes, respectively.

The P.E.D values indicate that the CrO_2 twisting modes are strongly coupled with vibrations of the same CrO_2 group and NO_2 groups with the 6-31G* basis set. In this compound, this mode is assigned to 152 cm^{-1}.

3.7.4 Nitrate Groups as Rings of Four Members

Also, in the bidentate type the calculations were performed with the three basis sets considering the two nitrate groups as a ring of four members where the deformations and torsion coordinates of these groups have been defined as proposed by Fogarasi et al. [40] and are observed in Table 3.13. The frequencies, IR intensities, Raman activities, and potential energy distribution obtained by B3LYP/6-31G*, B3LYP/Lanl2DZ, and B3LYP/6-311 + G calculations appear in Tables 3.7, 3.14, and 3.15, respectively. In all cases, the comparison of the the-oretical values with the respective experimental values are observed in the respective Tables. Although the best results are obtained with a B3LYP/Lanl2DZ calculation with a RMSD final equal at 12.5 cm^{-1}, the assignment was performed with B3LYP/6-31G* method because the vibrational modes appear more defined. In this case, a notable change in the assignment, in relation to the above bidentate considerations, is observed.

3.7.5 Nitrate Group

As it is shown in Table 3.7, the assignment of the two N=O and NO_2 antisym-metric stretchings does not change in reference to the above bidentate type. It is possible to observe some differences only in the frequencies of the two NO_2 symmetric modes. In this case, these modes appear at higher frequencies than the NO_2 deformation modes and the two N=O out-of-plane deformation modes are observed at lower frequencies. These last modes, are assigned to the same fre-quencies as in the above case while the N=O torsion and two NO_2 torsion modes (out-of-phase and in-phase) are observed in the lower frequency region, as shown in Table 3.7.

3.7.6 Chromyl Group

For this group, the calculations predict the two Cr=O stretchings, the CrO_2 bending, and the four Cr–O stretching modes at the same frequencies as the above bidentate coordination The only change is calculated for the wagging, rocking, and twisting modes. In the lower region, the A modes are calculated with the two basis sets strongly mixed while in the two bidentate coordinations of the B modes are

Table 3.13 Definition of natural internal coordinates for chromyl nitrate with bidentate coordination adopted for nitrate groups (as two rings of four members)

Symmetry A	
$S_1 = s\ (5\text{-}7) + s\ (9\text{-}11)$	v (N=O) ip
$S_2 = s\ (4\text{-}5) + s\ (8\text{-}9) - s\ (5\text{-}6) - s\ (9\text{-}10)$	va (NO$_2$) ip
$S_3 = q\ (1\text{-}2) + q\ (1\text{-}3)$	vs (Cr=O)
$S_4 = s\ (4\text{-}5) + s\ (5\text{-}6) + s\ (8\text{-}9) + s\ (9\text{-}10)$	vs (NO$_2$) ip
$S_5 = \beta\ (6\text{-}5\text{-}4) + \beta\ (4\text{-}1\text{-}6) + \beta\ (8\text{-}9\text{-}10) + \beta\ (10\text{-}1\text{-}8) - \beta\ (5\text{-}4\text{-}1) - \beta$ $(1\text{-}6\text{-}5) - \beta\ (9\text{-}10\text{-}1) - \beta\ (1\text{-}8\text{-}9)$	δ (NO$_2$) ip
$S_6 = \gamma\ (11\text{-}9\text{-}8\text{-}10) + \gamma\ (7\text{-}5\text{-}4\text{-}6)$	γ N=O ip
$S_7 = \beta\ (4\text{-}5\text{-}7) + \beta\ (8\text{-}9\text{-}11) - \beta\ (5\text{-}6\text{-}7) - \beta\ (10\text{-}9\text{-}11)$	δ (O=N–O) ip
$S_8 = \theta\ (2\text{-}1\text{-}3)$	δ (CrO$_2$)
$S_9 = r\ (1\text{-}4) + r\ (1\text{-}6) + r\ (1\text{-}8) + r\ (1\text{-}10)$	vs (Cr–O)
$S_{10} = \psi\ (2\text{-}1\text{-}6) + \psi\ (2\text{-}1\text{-}4) + \psi\ (3\text{-}1\text{-}8) + \psi\ (3\text{-}1\text{-}10) - \psi\ (3\text{-}1\text{-}6)$ $- \psi\ (3\text{-}1\text{-}4) - \psi\ (2\text{-}1\text{-}8) - \psi\ (2\text{-}1\text{-}10)$	ρ (CrO$_2$)
$S_{11} = r\ (1\text{-}4) + r\ (1\text{-}8) - r\ (1\text{-}6) - r\ (1\text{-}10)$	va (Cr–O)
$S_{12} = \phi\ (8\text{-}1\text{-}4) - \phi\ (10\text{-}1\text{-}6)$	δa (O–Cr–O)
$S_{13} = \tau\ (6\text{-}1\text{-}4\text{-}5) + \tau\ (4\text{-}5\text{-}6\text{-}1) + \tau\ (8\text{-}1\text{-}10\text{-}9) + \tau\ (10\text{-}9\text{-}8\text{-}1) - \tau\ (1\text{-}4\text{-}5\text{-}6)$ $+ \tau\ (5\text{-}6\text{-}1\text{-}4) + \tau\ (1\text{-}10\text{-}9\text{-}8) - \tau\ (9\text{-}8\text{-}1\text{-}10)$	τ (NO$_2$) ip
$S_{14} = \tau\ (6\text{-}1\text{-}4\text{-}5) + \tau\ (4\text{-}5\text{-}6\text{-}1) + \tau\ (1\text{-}10\text{-}9\text{-}8) + \tau\ (9\text{-}8\text{-}1\text{-}10) - \tau\ (1\text{-}4\text{-}5\text{-}6)$ $- \tau\ (5\text{-}6\text{-}1\text{-}4) - \tau\ (8\text{-}1\text{-}10\text{-}9) - \tau\ (10\text{-}9\text{-}8\text{-}1)$	τ (NO$_2$) op
Symmetry B	
$S_{15} = s\ (5\text{-}7) - s\ (9\text{-}11)$	v (N=O) op
$S_{16} = s\ (4\text{-}5) + s\ (9\text{-}10) - s\ (5\text{-}6) - s\ (8\text{-}9)$	va (NO$_2$) op
$S_{17} = q\ (1\text{-}2) - q\ (1\text{-}3)$	va (Cr=O)
$S_{18} = s\ (4\text{-}5) + s\ (5\text{-}6) - s\ (8\text{-}9) - s\ (9\text{-}10)$	vs (NO$_2$) op
$S_{19} = \beta\ (6\text{-}5\text{-}4) + \beta\ (4\text{-}1\text{-}6) + \beta\ (9\text{-}10\text{-}1) + \beta\ (1\text{-}8\text{-}9) - \beta\ (5\text{-}4\text{-}1) - \beta\ (1\text{-}6\text{-}5)$ $- \beta\ (8\text{-}9\text{-}10) - \beta\ (10\text{-}1\text{-}8)$	δ (NO$_2$) op
$S_{20} = \gamma\ (11\text{-}9\text{-}8\text{-}10) - \gamma\ (7\text{-}5\text{-}4\text{-}6)$	γ N=O op
$S_{21} = \beta\ (4\text{-}5\text{-}7) + \beta\ (10\text{-}9\text{-}11) - \beta\ (5\text{-}6\text{-}7) - \beta\ (8\text{-}9\text{-}11)$	δ (O=N–O) op
$S_{22} = r\ (1\text{-}4) + r\ (1\text{-}6) - r\ (1\text{-}8) - r\ (1\text{-}10)$	vs (Cr ← O)
$S_{23} = \psi\ (2\text{-}1\text{-}6) + \psi\ (3\text{-}1\text{-}4) + \psi\ (2\text{-}1\text{-}4) + \psi\ (3\text{-}1\text{-}6) - \psi\ (2\text{-}1\text{-}8) - \psi$ $(3\text{-}1\text{-}10) - \psi\ (2\text{-}1\text{-}10) - \psi\ (3\text{-}1\text{-}8)$	wag (CrO$_2$)
$S_{24} = \psi\ (2\text{-}1\text{-}6) + \psi\ (3\text{-}1\text{-}6) + \psi\ (2\text{-}1\text{-}8) + \psi\ (3\text{-}1\text{-}8) - \psi\ (2\text{-}1\text{-}4) - \psi$ $(3\text{-}1\text{-}4) - \psi\ (2\text{-}1\text{-}10) - \psi\ (3\text{-}1\text{-}10)$	τw (CrO$_2$)
$S_{25} = \tau\ (2\text{-}1\text{-}4\text{-}5) + \tau\ (3\text{-}1\text{-}4\text{-}5) + \tau\ (2\text{-}1\text{-}8\text{-}9) + \tau\ (3\text{-}1\text{-}8\text{-}9) + \tau\ (2\text{-}1\text{-}10\text{-}9)$ $+ \tau\ (3\text{-}1\text{-}10\text{-}9) + \tau\ (2\text{-}1\text{-}4\text{-}5) + \tau\ (3\text{-}1\text{-}4\text{-}5)$	τ (N=O)
$S_{26} = \phi\ (8\text{-}1\text{-}4) + \phi\ (10\text{-}1\text{-}6)$	δs (O–Cr–O)
$S_{27} = r\ (1\text{-}6) + r\ (1\text{-}8) - r\ (1\text{-}4) - r\ (1\text{-}10)$	va (Cr ← O)

Q Cr=O bond distance; r Cr–O bond distance; s N–O bond distance; θ O=Cr=O bond angle; ϕ O–Cr–O bond angle; ψ O=Cr–O bond angles; β O–N–O bond angle
v stretching; δ deformation; ρ in the plane bending or rocking; γ out of plane bending or wagging; τw twisting; a antisymmetric; s symmetric; ip in phase; op out of phase
Reprinted from [47], Copyright 2008, with permission from Elsevier

Table 3.14 Observed and calculated wavenumbers (cm^{-1}), potential energy distribution, and assignment for bidentate chromyl nitrate considering the nitrate groups as ring of four members

Mode	Observed[a]	Calculated[b]	SQM[c]	IR int.[d]	Raman act.[e]	PED (≥ 10 %)
Modes A symmetry						
1	1642	1560	1646	180.0	37.0	82 S_1
2	1216	1117	1207	138.0	21.0	61 S_2 + 33 S_4
3	959	1076	953	60.4	43.6	95 S_3
4	835	869	823	129.3	10.9	73 S_{20} + 20 S_4
5	781	712	794	32.5	5.1	77 S_9 + 11 S_{14}
6	774	682	780	0.9	0.4	57 S_5 + 20 S_9 + 12 S_{20}
7	686	618	665	1.2	7.5	37 S_2 + 27 S_7 + 22 S_4
8	460	450	452	3.6	15.5	61 S_8 + 23 S_9
9	349	345	355	3.7	14.4	42 S_8 + 25 S_9 + 12 S_5 + 10 S_{11}
10	242	255	252	0.0	2.5	34 S_{10} + 24 S_{26} + 16 S_{14} + 16 S_{25}
11	206	191	207	1.2	1.8	56 S_{11} + 33 S_9
12	206	183	180	0.0	2.5	79 S_{25} + 13 S_{14}
13	152	144	143	0.1	3.1	45 S_{10} + 32 S_{26}
14	70	70	68	0.0	8.0	51 S_{14} + 27 S_{25} + 19 S_{26}
Modes B symmetry						
15	1614	1542	1631	687.0	13.3	82 S_{15}
16	1232	1126	1216	215.0	0.9	60 S_{16} + 33S_{18}
17	945	1066	951	56.1	32.3	95 S_{17}
18	800	861	819	77.4	0.5	78 S_6 + 23 S_{18}
19	781	710	782	91.9	2.3	47 S_9 + 33 S_{19}
20	774	681	773	20.8	0.6	49 S_6 + 32 S_{19}
21	686	617	658	4.3	1.9	40 S_{16} + 25 S_{18} + 22 S_{21}
22	446	431	450	68.7	9.1	39 S_{22} + 15 S_{23} + 14 S_{19} + 13 S_{21} + 12 S_{27}
23	273	281	279	11.3	2.5	61 S_{23} + 15 S_{12} + 11 S_{13}
24	223	225	229	4.0	0.7	52 S_{24} + 22 S_{23}
25	152	173	171	1.0	0.9	34 S_{27} + 25 S_{22} + 17 S_{23} + 13 S_{13}
26	100	97	96	0.3	2.5	39 S_{24} + 24 S_{13} + 14 S_{23} + 12 S_{27}
27	100	87	89	9.9	0.1	42 S_{27} + 24 S_{22} + 21 S_{24}
RMSD (cm^{-1})		62.2	12.5			

[a] This work
[b] DFT B3LYP/Lanl2DZ
[c] From scaled quantum mechanics force field
[d] Units are km mol^{-1}
[e] Raman activities in Å^4 (amu)$^{-1}$

calculated to be coupled between them. For this analysis, the two bidentate coordinations are possible but, not the monodentate type because the Cr–O symmetric stretching modes should be observed with higher intensity than the CrO_2 bending mode. Moreover, as we will observe next, the force constants of the Cr–O stretchings cannot be bigger than the corresponding to Cr=O stretchings.

Table 3.15 Observed and calculated wavenumbers (cm^{-1}), potential energy distribution, and assignment for bidentate chromyl nitrate considering the nitrate groups as ring of four members

Mode	Observed[a]	Calculated[b]	SQM[c]	IR int.[d]	Raman act.[e]	PED ($\geq$10 %)
Modes A symmetry						
1	1642	1565	1637	207.3	40.0	78 S_1 + 11 S_4
2	1216	1123	1168	129.5	19.6	51 S_2 + 40 S_7
3	959	1051	952	78.4	45.2	95 S_3
4	835	863	879	107.8	9.9	71 S_4 + 21 S_7
5	781	728	808	35.6	2.7	70 S_5 + 19 S_9
6	774	714	774	0.0	0.3	68 S_6 + 16 S_{20} + 11 S_{13}
7	686	624	655	3.3	6.0	49 S_2 + 23 S_7 + 17 S_4
8	460	447	459	4.1	16.3	67 S_8 + 21 S_9
9	349	339	349	4.1	13.6	35 S_8 + 30 S_9 + 12 S_{11} + 12 S_5
10	242	257	259	0.0	1.8	38 S_{10} + 19 S_{26} + 17 S_{25} + 17 S_{14}
11	206	201	202	1.9	1.5	50 S_{11} + 33 S_9
12	152	173	173	0.9	2.4	31 S_{27} + 26 S_{22} + 16 S_{23} + 11 S_{12}
13	100	96	99	14.9	0.2	51 S_{13} + 16 S_{25} + 14 S_{12} + 10 S_{24}
14	76	67	69	0.1	7.5	48 S_{14} + 30 S_{26} + 19 S_{25}
Modes B symmetry						
15	1614	1547	1619	193.3	0.7	79 S_{15} + 11 S_{18}
16	1232	1125	1174	102.1	29.5	50 S_{16} + 40 S_{21}
17	945	1045	952	53.4	0.4	94 S_{17}
18	800	858	878	60.0	2.4	68 S_{18} + 22 S_{21}
19	781	727	804	42.7	1.0	72 S_{19} + 17 S_{22}
20	774	714	773	6.8	1.1	66 S_{20} + 15 S_{10} + 11 S_{14}
21	686	621	646	77.7	6.5	52 S_4 + 19 S_{18} + 19 S_{21}
22	446	437	452	7.7	2.3	37 S_{22} + 17 S_{23} + 15 S_{19} + 13 S_{27} + 11 S_{21}
23	273	273	276	3.5	0.7	60 S_{23} + 12 S_{12} + 11 S_{13}
24	223	226	227	2.1	1.1	50 S_{24} + 17 S_{23}
25	206	190	193	0.6	3.2	35 S_{25} + 33 S_{11} + 18 S_9
26	152	141	143	0.0	3.1	46 S_{10} + 28 S_{26} + 14 S_{14}
27	100	81	79	193.3	0.7	41 S_{27} + 22 S_{22} + 20 S_{24}
RMSD (cm^{-1})		53.0	26.6			

[a] This work

[b] DFT B3LYP/6-311+G

[c] From scaled quantum mechanics force field

[d] Units are km mol^{-1}

[e] Raman activities in Å^4 (amu)$^{-1}$

3.7.7 Force Field

Having a secure assignment for the experimentally studied chromyl nitrate, the corresponding force constants were estimated using the scaling procedure of Pulay et al. [41], as mentioned before. The harmonic force fields in cartesian coordinates were transformed to the local symmetry or "natural" coordinates proposed by Fogarasi et al. [40], as defined in Tables 3.13, 3.14, 3.15, 3.16, and 3.17 (See Figs. 3.3 and 3.4) considering in the first case the mode of coordination adopted by nitrate groups as monodentate and in the two following cases as bidentate. The scaling factors affecting the main force constants were subsequently calculated by an iterative procedure [42, 43] to have the best possible fit between observed and theoretical frequencies. The resulting numbers for the three cases considered are collected in Table 3.18. These values are quite satisfactory, considering that the experimental frequencies were not corrected for anharmonicity. The frequencies, IR intensities, Raman activities, and potential energy distribution obtained for chromyl nitrate appear together with the values reached for the corresponding RMSD values in Tables 3.7, 3.9, and 3.10 and Tables 3.11, 3.12, 3.14, and 3.15, for the three basis sets and for the three coordination modes considered for the nitrate groups in the compound. The force constants appearing in Table 3.19 expressed in terms of simple valence internal coordinates were calculated from the corresponding scaled force fields by using the expression: $F_i = U^t F_s U$, where F_i is the force constant matrix in terms of simple valence internal coordinates, F_s is the force constant matrix in terms of natural coordinates, U is the orthogonal matrix relating the natural coordinates to the simple valence internal coordinates, and U^t is the transposed matrix of the U matrix.

It is interesting to compare the principal force constants calculated at the B3LYP/Lanl2DZ, B3LYP/6-31G*, and B3LYP/6-311++G levels for the common vibrations, which were collected in Table 3.19. In general, the calculated force constant values for the two bidentate coordinations considered here, with the B3LYP/Lanl2DZ method are approximately the same as the calculated by B3LYP/6-311++G calculation. Obviously, some force constant values vary when the coordination mode of the nitrate group changes. As expected, the force constants of the N=O and Cr–O stretching change with the coordination mode of the nitrate group, being greater in the monodentate coordination than in the bidentate coordination. The force constants of Cr=O and O–Cr–O deformation modes are practically the same in the two cases, while other modes change as the coordination modes of the nitrate groups change. In CrO_2F_2 and CrO_2Cl_2, the scaled GVFF force constants (B3LYP/Lanl2DZ method) for the Cr=O stretchings are 7.443 and 7.122 mdyn $Å^{-1}$, respectively, while the corresponding force constants for the O=Cr=O deformations are 1.110 and 0.938 mdyn $Å$ rad^{-2} [18]. This difference in the force constant values in reference to chromyl nitrate cannot be attributed to geometric parameters because they are practically the same in the three compounds. The corresponding values are lower in CrO_2F_2 and CrO_2Cl_2 because the scaled Cr=O frequencies (1015 and 990 cm^{-1}) are higher than the

Table 3.16 Definition of natural internal coordinates for chromyl nitrate with monodentate coordination adopted by nitrate groups

Symmetry A

$S_1 = s$ (5-7) $+ s$ (5-6) $+ s$ (9-11) $+ s$ (9-10)	vs (NO$_2$) ip
$S_2 = s$ (5-7) $+ s$ (9-11) $- s$ (5-6) $- s$ (9-10)	va (NO$_2$) ip
$S_3 = q$ (1-2) $+ q$ (1-3)	vs (Cr=O)
$S_4 = 2\beta$ (6-5-7) $+ 2\beta$ (10-9-11) $- \beta$ (4-5-6) $- \beta$ (4-5-7) $- \beta$ (8-9-10) $- \beta$ (8-9-11)	δ (NO$_2$) ip
$S_5 = \gamma$ (11-9-8-10) $- \gamma$ (7-5-4-6)	γ N=O op
$S_6 = 2\beta$ (4-5-6) $+ 2\beta$ (8-9-10) $- \beta$ (4-5-7) $- \beta$ (6-5-7) $- \beta$ (8-9-11) $- \beta$ (10-9-11)	δ (O=N–O) ip
$S_7 = s$ (5-4) $+ s$ (9-8)	vs (N–O)
$S_8 = r$ (1-4) $- r$ (1-8)	va (Cr–O)
$S_9 = r$ (1-4) $+ r$ (1-8)	vs (Cr–O)
$S_{10} = \tau$ (10-9-8-1) $+ \tau$ (11-9-8-1) $+ \tau$ (7-5-4-1) $+ \tau$ (6-5-4-1)	τ (NO$_2$) ip
$S_{11} = \alpha$ (5-4-1) $+ \alpha$ (9-8-1)	δ (N–O–Cr) ip
$S_{12} = \tau$ (3-1-4-5) $+ \tau$ (2-1-8-9)	ρ (NO$_2$) ip
$S_{13} = \psi$ (2-1-8) $+ \psi$ (3-1-4) $- \psi$ (2-1-4) $- \psi$ (3-1-8)	ρ (CrO$_2$)
$S_{14} = \phi$ (4-1-8)	δ (O–Cr–O)

Symmetry B

$S_{15} = s$ (5-7) $+ s$ (5-6) $- s$ (9-11) $- s$ (9-10)	vs (NO$_2$) op
$S_{16} = s$ (5-7) $+ s$ (9-10) $- s$ (5-6) $- s$ (9-11)	va (NO$_2$) op
$S_{17} = q$ (1-2) $- q$ (1-3)	va (Cr=O)
$S_{18} = 2\beta$ (6-5-7) $- 2\beta$ (10-9-11) $- \beta$ (4-5-6) $- \beta$ (4-5-7) $+ \beta$ (8-9-10) $+ \beta$ (8-9-11)	δ (NO$_2$) op
$S_{19} = \gamma$ (11-9-8-10) $+ \gamma$ (7-5-4-6)	γ N=O ip
$S_{20} = 2\beta$ (4-5-6) $- 2\beta$ (8-9-10) $+ \beta$ (8-9-11) $+ \beta$ (10-9-11) $- \beta$ (4-5-7) $- \beta$ (6-5-7)	δ (O=N–O) op
$S_{21} = s$ (5-4) $- s$ (9-8)	va (N–O)
$S_{22} = \theta$ (2-1-3)	δ (CrO$_2$)
$S_{23} = \psi$ (2-1-4) $+ \psi$ (3-1-4) $- \psi$ (2-1-8) $- \psi$ (3-1-8)	wag (CrO$_2$)
$S_{24} = \alpha$ (5-4-1) $- \alpha$ (9-8-1)	δ (N–O–Cr) op
$S_{25} = \tau$ (3-1-4-5) $- \tau$ (2-1-8-9)	ρ (NO$_2$) op
$S_{26} = \tau$ (10-9-8-1) $+ \tau$ (11-9-8-1) $- \tau$ (7-5-4-1) $- \tau$ (6-5-4-1)	τ (NO$_2$) op
$S_{27} = \psi$ (3-1-4) $+ \psi$ (3-1-8) $- \psi$ (2-1-4) $- \psi$ (2-1-8)	τwis (CrO$_2$)

q Cr=O bond distance; r Cr–O bond distance; s N–O bond distance; θ O=Cr=O bond angle; ϕ O–Cr–O bond angle; ψ O=Cr–O bond angles; α Cr–O–N bond angle; β O–N–O bond angle; v stretching; δ deformation; ρ in the plane bending or rocking; γ out of plane bending or wagging; τw twisting; a antisymmetric, s symmetric; ip in phase; op out of phase
Reprinted from [47], Copyright 2008, with permission from Elsevier

values for CrO$_2$(NO$_3$)$_2$ as shown in Tables 3.7, 3.8, and 3.9. The lower values of the force constants of O=Cr=O deformations in CrO$_2$F$_2$ and CrO$_2$Cl$_2$ in comparison with CrO$_2$(NO$_3$)$_2$ can also be attributed to the scaled O=Cr=O frequencies that are higher in this last compound.

The force constants of Cr–O stretchings considering bidentate nitrate groups are near the expected value reported by Hester et al. [44] for an M–O frequency

Table 3.17 Definition of natural internal coordinates for chromyl nitrate with bidentate coordination adopted by nitrate groups

Symmetry A	
$S_1 = s\ (5\text{-}7) + s\ (9\text{-}11)$	ν (N=O) ip
$S_2 = s\ (4\text{-}5) + s\ (9\text{-}10) - s\ (5\text{-}6) - s\ (8\text{-}9)$	νa (NO$_2$) op
$S_3 = q\ (1\text{-}2) + q\ (1\text{-}3)$	νs (Cr=O)
$S_4 = 2\beta\ (4\text{-}5\text{-}6) + 2\beta\ (8\text{-}9\text{-}10) - \beta\ (4\text{-}5\text{-}7) - \beta\ (6\text{-}5\text{-}7) - \beta\ (8\text{-}9\text{-}11)$ $\quad - \beta\ (10\text{-}9\text{-}11)$	δ (NO$_2$) ip
$S_5 = \gamma\ (11\text{-}9\text{-}8\text{-}10) - \gamma\ (7\text{-}5\text{-}4\text{-}6)$	γ N=O op
$S_6 = s\ (4\text{-}5) + s\ (5\text{-}6) - s\ (8\text{-}9) - s\ (9\text{-}10)$	νs (NO$_2$) op
$S_7 = 2\beta\ (5\text{-}6\text{-}7) - 2\beta\ (9\text{-}10\text{-}11) + \beta\ (8\text{-}9\text{-}10) + \beta\ (8\text{-}9\text{-}11) - \beta\ (4\text{-}5\text{-}6)$ $\quad - \beta\ (4\text{-}5\text{-}7)$	δ (O=N–O) op
$S_8 = \theta\ (2\text{-}1\text{-}3)$	δ (CrO$_2$)
$S_9 = r\ (1\text{-}4) + r\ (1\text{-}6) + r\ (1\text{-}8) + r\ (1\text{-}10)$	νs (Cr–O)
$S_{10} = \psi\ (2\text{-}1\text{-}6) + \psi\ (2\text{-}1\text{-}4) + \psi\ (3\text{-}1\text{-}8) + \psi\ (3\text{-}1\text{-}10) - \psi\ (3\text{-}1\text{-}6)$ $\quad - \psi\ (3\text{-}1\text{-}4) - \psi\ (2\text{-}1\text{-}8) - \psi\ (2\text{-}1\text{-}10)$	ρ (CrO$_2$)
$S_{11} = r\ (1\text{-}4) + r\ (1\text{-}8) - r\ (1\text{-}6) - r\ (1\text{-}10)$	νa (Cr–O)
$S_{12} = \psi\ (2\text{-}1\text{-}6) + \psi\ (3\text{-}1\text{-}6) + \psi\ (2\text{-}1\text{-}8) + \psi\ (3\text{-}1\text{-}8) - \psi\ (2\text{-}1\text{-}4)$ $\quad - \psi\ (3\text{-}1\text{-}4) - \psi\ (2\text{-}1\text{-}10) - \psi\ (3\text{-}1\text{-}10)$	τw (CrO$_2$)
$S_{13} = r\ (1\text{-}4) + r\ (1\text{-}6) - r\ (1\text{-}8) - r\ (1\text{-}10)$	νs (Cr ← O)
$S_{14} = \tau\ (10\text{-}9\text{-}8\text{-}1) + \tau\ (11\text{-}9\text{-}8\text{-}1) + \tau\ (7\text{-}5\text{-}4\text{-}1) + \tau\ (6\text{-}5\text{-}4\text{-}1)$	τ (NO$_2$) ip
Symmetry B	
$S_{15} = s\ (5\text{-}7) + s\ (9\text{-}11)$	ν (N=O) op
$S_{16} = s\ (4\text{-}5) + s\ (8\text{-}9) - s\ (5\text{-}6) - s\ (9\text{-}10)$	νa (NO$_2$) ip
$S_{17} = q\ (1\text{-}2) - q\ (1\text{-}3)$	νa (Cr=O)
$S_{18} = 2\beta\ (4\text{-}5\text{-}6) - 2\beta\ (8\text{-}9\text{-}10) + \beta\ (8\text{-}9\text{-}10) + \beta\ (10\text{-}9\text{-}11) - \beta\ (4\text{-}5\text{-}7)$ $\quad - \beta\ (6\text{-}5\text{-}7)$	δ (NO$_2$) op
$S_{19} = \gamma\ (11\text{-}9\text{-}8\text{-}10) + \gamma\ (7\text{-}5\text{-}4\text{-}6)$	γ N=O ip
$S_{20} = s\ (4\text{-}5) + s\ (5\text{-}6) + s\ (8\text{-}9) + s\ (9\text{-}10)$	νs (NO$_2$) ip
$S_{21} = 2\beta\ (5\text{-}6\text{-}7) + 2\beta\ (9\text{-}10\text{-}11) - \beta\ (4\text{-}5\text{-}6) - \beta\ (4\text{-}5\text{-}7) - \beta\ (8\text{-}9\text{-}10)$ $\quad - \beta\ (8\text{-}9\text{-}11)$	δ (O=N–O) ip
$S_{22} = = r\ (1\text{-}4) + r\ (1\text{-}6) - r\ (1\text{-}8) - r\ (1\text{-}10)$	νa (Cr ← O)
$S_{23} = \tau\ (10\text{-}9\text{-}8\text{-}1) + \tau\ (11\text{-}9\text{-}8\text{-}1) - \tau\ (7\text{-}5\text{-}4\text{-}1) - \tau\ (6\text{-}5\text{-}4\text{-}1)$	τ (NO$_2$) op
$S_{24} = \psi\ (2\text{-}1\text{-}6) + \psi\ (3\text{-}1\text{-}4) + \psi\ (2\text{-}1\text{-}4) + \psi\ (3\text{-}1\text{-}6) - \psi\ (2\text{-}1\text{-}8)$ $\quad - \psi\ (3\text{-}1\text{-}10) - \psi\ (2\text{-}1\text{-}10) - \psi\ (3\text{-}1\text{-}8)$	wag (CrO$_2$)
$S_{25} = \phi\ (8\text{-}1\text{-}4) + \phi\ (10\text{-}1\text{-}6)$	δs (O–Cr–O)
$S_{26} = \phi\ (8\text{-}1\text{-}4) - \phi\ (10\text{-}1\text{-}6)$	δa (O–Cr–O)
$S_{27} = \tau\ (3\text{-}1\text{-}4\text{-}5) + \tau\ (3\text{-}1\text{-}10\text{-}9) - \tau\ (2\text{-}1\text{-}8\text{-}9) - \tau\ (2\text{-}1\text{-}6\text{-}5)$	τ (N=O)

q Cr=O bond distance; r Cr–O bond distance; s N–O bond distance; θ O=Cr=O bond angle; ϕ O–Cr–O bond angle; ψ O=Cr–O bond angles; β O–N–O bond angle; ν stretching; δ deformation; ρ in the plane bending or rocking; γ out of plane bending or wagging; τw twisting; a antisymmetric; s symmetric; ip in phase; op out of phase

Reprinted from [47], Copyright 2008, with permission from Elsevier

(321 cm^{-1} in bidentate nitrate; 2.0 mdyn Å^{-1}). The greater value for this force constant (about 7.82 mdyn Å^{-1}) in the monodentate type suggests that the monodentate coordination for the nitrate groups in chromyl nitrate is impossible; therefore, this compound would have multiple coordination.

Table 3.18 Scale factors for the force field of chromyl nitrate

Coordinates	[a]Chromyl nitrate					
	Monodentate			Bidentate		
	Lanl2DZ	6-31G*	6-311++G	Lanl2DZ	6-31G*	6-311++G
ν (N=O)	1.702	0.641	0.825	1.147	0.899	1.137
ν (N–O)	0.798	0.641	0.825	1.147	0.899	1.137
ν (Cr=O)	0.774	0.730	0.815	0.774	0.728	0.816
ν (Cr–O)	1.406	1.029	1.426	1.118	0.908	1.061
δ (O=N=O)	1.090	0.968	1.055			
δ (O–N–O)				1.255	1.041	1.248
δ (O=N–O)	1.090	0.968	1.055	1.255	1.041	1.248
δ (O=Cr=O)	1.117	0.917	0.942	1.063	0.941	1.083
δ (O–Cr–O)	1.117	0.917	0.942	1.063	0.941	1.083
δ (O–N–Cr)	1.117	0.917	0.942			
τ (O–N–O)	1.034	1.006	1.054	0.984	0.988	0.999
γ (N=O)	1.302	1.002	1.185	1.254	0.999	1.159
ρ (O–Cr–O)	0.861	0.791	0.913	0.785	0.854	0.825
Wagg (O–Cr–O)	0.861	0.791	0.913	0.785	0.854	0.825
τw (O–Cr–O)	0.861	0.791	0.913	0.785	0.988	0.999

ν stretching; δ deformation; ρ rocking; wag (γ) wagging; τw torsion
[a] This work
Reprinted from [47], Copyright 2008, with permission from Elsevier

When the nitrate coordination is monodentate the force constants of N=O stretchings are higher than other coordination modes while in the bidentate type the force constants of the N–O stretchings are higher than the monodentate type. Moreover, our values for the force constants N=O stretching are in agreement with that the reported value of 11.83 mdyn Å^{-1} for N_2O compound [8] while it is different from the 14.51 mdyn Å^{-1} value reported for the N_2O_2 compound [8]. The structures of both compounds are different from chromyl nitrate, being N_2O linear and N_2O_2 angular with a 90° O–N–N angle. The force constant values reported for KNO_3 by Beattie et al. [39] were: 9.26 mdyn Å^{-1} for N=O stretching, 6.72 mdyn Å^{-1} for N–O stretching, 1.54 mdyn Å rad^{-2} for O–N–O deformation, and 1.54 mdyn Å rad^{-2} for O=N–O deformation. The force constant values reported by Brintzinger et al. [45] for the free anion (6.35 mdyn Å^{-1} for N–O stretching, 2.05 mdyn Å^{-1} for N–O/N–O stretching, and 0.54 mdyn Å rad^{-2} for O–N–O deformation) are near the cited values by Topping for D_{3h} nitrate ion (6.5 mdyn Å^{-1} for N–O stretching, 2.05 mdyn Å^{-1} for N–O/N–O stretching, and 0.54 mdyn Å rad^{-2} for O–N–O deformation) [46]. For chromyl nitrate, those force constant values for the bidentate type are near to monodentate coordination as can be seen in Table 3.19. The observed differences in the force constants for KNO_3 can be attributed to the calculations because in that compound they were carried out using three observed N–O stretching frequencies (1,460, 1,293, and 1,031 cm^{-1}) and the C_{2v} bidentate model. The interaction force constants N=O/N–O for the monodentate type in chromyl nitrate are slightly higher than the 1.11 mdyn Å^{-1} value

Table 3.19 Comparison of scaled internal force constants for chromyl nitrate

Coordinates [a]Chromyl nitrate

	Monodentate			Bidentate			[#]Bidentate		
	Lanl2DZ	6-31G*	6-311++G	Lanl2DZ	6-31G*	6-311++G	Lanl2DZ	6-31G*	6-311++G
f (N=O)	16.18	16.25	15.83	11.74	11.44	11.71	11.55	11.62	11.07
f (N–O)	3.19	3.38	3.22	4.62	4.96	4.62	4.96	4.96	4.62
f (Cr=O)	6.55	6.55	6.56	6.55	6.53	6.57	6.55	6.57	6.56
f (Cr–O)	7.82	6.09	7.34	1.64	1.44	1.51	1.34	1.38	1.27
f (O=N=O)	1.57	1.62	1.58	–	–	–	–	–	–
f (O=N–O)	2.05	2.24	2.09	1.83	1.74	1.87	1.49	1.32	1.57
f (O=Cr=O)	2.31	2.53	2.26	1.73	1.66	1.74	1.58	1.62	1.63
f (O–Cr–O)	0.86	0.80	0.74	0.91	0.93	0.93	0.75	0.65	0.66
f (O–N–Cr)	1.96	2.41	1.96	–	–	–	–	–	–
f (N=O)/(N–O)	1.74	1.81	1.71	1.33	1.29	1.25	1.66	1.95	1.60
f (N=O)/(Cr–O)	1.16	1.16	1.16	−0.12	−0.07	−0.11	−0.27	−0.47	−0.39
f (N–O)/(N–O)	2.00	2.06	2.09	−1.30	−1.29	−1.29	−1.35	−1.61	−1.36

Units are mdyn Å^{-1} for stretching and stretching/stretching interaction and mdyn Å rad^{-2} for angle deformations

[a] This work

[#] Considering the nitrate groups as ring of four members

Reprinted from [47], Copyright 2008, with permission from Elsevier

reported for bidentate KNO_3 [44] but close to the obtained values considering the nitrate groups in chromyl nitrate as bidentate.

The analysis of the force constants suggests that the coordination that better represents the group nitrate in chromyl nitrate is the bidentate because the obtained values for this case agree with the one reported by the literature values for this coordination mode.

3.8 Conclusions

In the present chapter, an approximate normal coordinate analysis, considering the mode of coordination adopted by nitrate groups as monodentate and bidentate, was proposed for chromyl nitrate.

The assignments previously made [1] were corrected and completed in accordance with the present theoretical results. The assignments of the 27 normal modes of vibration corresponding to chromyl nitrate are reported.

The method that best reproduces the experimental vibrational frequencies, considering the two coordination types of the nitrate groups for chromyl nitrate, it is the B3LYP/6-31G*.

The NBO and AIM analysis confirm the hexacoordination of the Cr atom in chromyl nitrate.

The Lanl2DZ, 6-31G*, and 6-311+G basis sets at the B3LYP level were employed for to obtain a molecular force field and vibrational frequencies.

An SQM force field was obtained for chromyl nitrate after adjusting the theoretically obtained force constants in order to minimize the difference between observed and calculated frequencies.

For the chromyl nitrate, a DFT molecular force field with the coordination mode adopted by nitrate groups as bidentate, computed using Lanl2DZ, 6-31G*, and 6-311+G basis sets are well represented.

Acknowledgments This work was subsidized with grants from CIUNT (Consejo de Investigaciones, Universidad Nacional de Tucumán), and CONICET (Consejo Nacional de Investigaciones Científicas y Técnicas, R. Argentina). The authors thank Prof. Tom Sundius for his permission to use MOLVIB.

References

1. E.L. Varetti, S.A. Brandán, A. Ben Altabef, Vib. Spectros. **5**, 219 (1993)
2. S.A. Brandán, A. Ben Altabef, E.L. Varetti, Spectrochim. Acta **51A**, 669 (1995)
3. M. Fernández Gómez, A. Navarro, S.A. Brandán, C. Socolsky, A. Ben Altabef, E.L. Varetti, J. Mol. Struct. (THEOCHEM) **626**, 101 (2003)
4. C. Socolsky, S.A. Brandán, A. Ben Altabef, E.L. Varetti, J. Mol Struct. (THEOCHEM) **672**, 45 (2004)
5. M.L. Roldán, H. Lanús, S.A. Brandán, J.J. López, E.L. Varetti, A. Ben Altabef, J. Argent. Chem. Soc. **92**, 53 (2004)
6. M.L. Roldán, S.A. Brandán, E.L. Varetti, A. Ben Altabef, Z. Anorg, Allg. Chem. **632**, 2495 (2006)
7. W.A. Guillory, M.L. Bernstein, J. Chem. Phys. **62**(3), 1059 (1975)
8. J. Laane y J. R. Ohlsen, Prog. Inorg. Chem., **27**, 465(1980)
9. B. Lippert, C.J.L. Lock, B. Rosenberg, M. Zvagulis, Inorg. Chem. **16**, 1525 (1977)
10. A. D. Harris, J. C. Trebellas, H. B. Jonassen, Inorg. Synth. **9**, 83(1967)
11. C.J. Marsden, K. Hedberg, M.M. Ludwig, G.L. Gard, Inorg. Chem. **30**, 4761 (1991)
12. M. Schmeisser, D. Lutzow, Angew. Chem. **66**, 230 (1954)
13. M. Schmeiser, Z. Angew, Chem. **67**(17–18), 493 (1955)
14. S.D. Brown, G.L. Gard, Inorg. Chem. **12**, 483 (1973)
15. G. Brauer (ed.), *Handbuch der Preparativen Anorganischen Chemie* (Enke, Stuttgart, 1975), p. 1523
16. W.H. Hartford, M. Darrin, Chem. Rev. **58**, 1 (1958)
17. J. C. Bailor, H. J. Emeléus, R. Myholm, A. F. Trotman-Dickenson, *Comprehensive Inorganic Chemistry*, Edit. (Pergamon Press, Oxford, 1975), p. 694
18. S. Bell, T.J. Dines, J. Phys. Chem. A **104**, 11403 (2000)
19. E. D. Glendening, A. E. Reed, J. E. Carpenter, F. Weinhold. NBO Version 3.1
20. R.F.W. Bader, *Atoms in Molecules, A Quantum Theory* (Oxford University Press, Oxford, 1990) ISBN 0198558651
21. R.J. Gillespie (ed.), *Molecular Geometry* (Van Nostrand-Reinhold, London, 1972)
22. R.J. Gillespie, I. Bytheway, T.H. Tang, R.F.W. Bader, Inorg. Chem. **35**, 3954 (1996)
23. AIM2000 designed by, University of Applied Sciences, Bielefeld, Germany

24. S. Wojtulewski, S.J. Grabowski, J. Mol. Struct. **621**, 285 (2003)
25. S.J. Grabowski, Monat. für Chem. **133**, 1373 (2002)
26. R.F.W. Bader, J. Phys. Chem. A **102**, 7314 (1998)
27. P.L.A. Popelier, J. Phys. Chem. A **102**, 1873 (1998)
28. U. Koch, P.L.A. Popelier, J. Phys. Chem. **99**, 9747 (1995)
29. G.L. Sosa, N. Peruchena, R.H. Contreras, E.A. Castro, J. Mol. Struct. (THEOCHEM) **401**, 77 (1997)
30. G.L. Sosa, N. Peruchena, R.H. Contreras, E.A. Castro, J. Mol. Struct. (THEOCHEM) **577**, 219 (2002)
31. S. A. Brandán, Estudio espectroscópico de Compuestos Inorgánicos Derivados de Metales de Transición, Doctoral Thesis, National University of Tucumán, R. Argentina, 1997
32. H. Siebert, *Anwendungen der schwingungsspektroskopie in der Anorganische Chemie*, (Springer, Berlin, 1966), p. 72
33. J.R. Ferraro, A. Walker, J. Chem. Phys. **42**, 1273 (1965)
34. J.R. Ferraro, A. Walker, J. Chem. Phys. **42**(4), 1278 (1965)
35. J. R. Ferraro y A. Walker, J. Chem. Phys. **45**(2), 550 (1967)
36. K. Nakamoto, *Infrared and Raman Spectra of Inorganic and Coordination Compounds*, 5th edn. (Wiley Inc., New York, 1997)
37. B.O. Field, C.J. Hardy, Quart. Rev. **18**, 361 (1964)
38. M. Chaabouni, T. Chausse, J.L. Pascal, J. Potier, J. Chem. Research **5**, 72 (1980)
39. I.R. Beattie, J.S. Ogden, D.D. Price, J. Chem. Soc. Dalton **10**, 1460 (1979)
40. G. Fogarasi, P. Pulay, in *Vibrational Spectra and Structure*, ed. by J. E. Durig, vol. 14 (Elsevier, Amsterdam, 1985), p. 125
41. P. Pulay, G. Fogarasi, F. Pang, J.E. Boggs, J. Am. Chem. Soc. **101**(10), 2550 (1979)
42. T. Sundius, J. Mol. Struct. **218**, 321 (1990)
43. T. Sundius, MOLVIB: a program for harmonic force field calculation, QCPE Program No. 604, (1991)
44. R.E. Hester, W.E.L. Grossman, Inorg. Chem. **5**, 1308 (1966)
45. H. Brintzinger, R.E. Hester, Inorg. Chem. 980 (1966)
46. G. Topping, Spectrochim. Acta **21**, 1743 (1965)
47. S.A. Brandán, M.L. Roldán, C. Socolsky, A. Ben Altabef, DFT Calculation of the Chromyl Nitrate, $CrO_2(NO_3)_2$: The Molecular Force Field, Spectrochim. Acta. A. Mol. Biomol. Spectrosc. **69**(3), 1027–1043 (2008)